DESIGN GRAND PRIX AWARD —

WORKS BY CREATIVE TAICHUNG TOP TEN DESIGNERS

ARTPOWER

Design Grand Prix Award — Works By Creative Taichung Top Ten Designers

Published by Artpower International Publishing Co., Ltd.

ARTPOWER™

Designer: Zheng Qin
Chief Editor: Mo Tingli

Address: Room C, 9/F., Sun House, 181 Des Voeux Road Central, Hong Kong, China
Tel: 852-31840676
Fax: 852-25432396

Editorial Department:
Address: G009, Floor 7th, Yimao Centre, Meiyuan Road, Luohu District, Shenzhen, China
Tel: 86-755-82913355
Fax: 86-755-82020029

Web: www.artpower.com.cn
E-mail: artpower@artpower.com.cn

ISBN 978-988-19973-2-6

Printed in China

During Global View Magazine's interview and in his book, Stanley Yen, the Godfather of tourism in Taiwan, mentioned "Not long ago, I received several friends from Hong Kong with Lung Ying-tai in Taichung, and many were professors, writers and chairs of parliaments. Through our arrangement, they found that Taichung was simply marvelous." The splendor, beauty, deliciousness and fun of Taichung make it the best tourism destination in central Taiwan. Convenient transportation, sophisticated living functions, a wide range of diverse, creatively designed restaurants, recreational activities and renowned nightlife culture have enriched Taichung's vibrant lifestyle. Consequently, Stanley Yen has praised Taichung's willingness to accept new ideas and propagate creativity. According to Common Wealth magazine's 2010 "Top 25 Happiness Municipality Survey", which was based on "Economy", "Environment", "Administration", "Culture & Education" and "Social Welfare", the analysis result indicated that Taichung citizens hold an optimistic outlook towards the development of their city, while their recognition and care about the city have also increased. In addition, did you know that the city's development is attributed to the endeavors of the creative maestros? Their masterpiece is inspired and nurtured by the magnificent city of Taichung. "Taichung is the place for young people to overturn tradition and attempt new creative ideas. Numerous Taiwanese creativities, such as chain bubble tea stores, boutique motels, luxury architecture originated from Taichung." (Reported by Common Wealth Magazine on June 29, 2011)

There are 12 universities in Taichung City that combine with the surrounding communities to form an arts and living creative district. After the elevation of the city of Taichung, the overall adjustment in the system will integrate the city with education and collaborate with the creative industry will be made so that job seekers, creative talents or education seekers can converge in Taichung, and thereby transform Taichung into a world class cultural creative industry with local characteristics. The ultimate objective is to develop a unique, innovative industry environment in Taichung. Greater Taichung's cultural convergence has culminated in the outstanding commercial spaces as Taichung's designers often pioneer new thinking and exports their incessant creativity globally and to China. The book intends to consolidate the creative essences of ten prominent design firms in Taichung that have either won major domestic and international design awards or are at the forefront of the commercial design industry in Taiwan, China, Southeast Asia or other countries. The book examines the firms closely from multiple perspectives, including their design provenance, process and creative philosophy in order to enable the general public to gain further understand about design. In addition, within the creative economy driven society today, the aim is to market Taichung through the designers' works. From the renowned Gongyi Road Shopping and Chingming District, Wenxing Fulfillment Amphitheatre to the Taichung Metropolitan Opera House, which is under construction, as well as the Taichung Gateway Park and the MRT system, they all demonstrate Taichung's resolve to become a pop culture capital. Furthermore, numerous surprising design and creative elements have been infused, such as arts, culture, music, dancing, opera, theatre, cinema and culinary culture etc. to not only foster the strength of the cultural industry, but also accumulate substantial economy of aesthetics.

In turn, this has generated Taichung's foreign exchange earnings and provides a large number of employment opportunities in the creative industry. Also, through the designers' innovativeness, the world can bear witness to Taichung's soft power and the value of its creative industry. Taichung may not be the capital of Taiwan, yet it will become the design hub of Taiwan. According to Prof. Liu YueHua from the Department of Urban Planning and Spatial Information, Feng Chia University "The memory of a city refers to the collective recollection of local residents. For instance, Taichung Train Station is an iconic memory of the people of Taichung, it is also in the collective memory and the relative coordinates for a majority of Taichung citizens. Its existence bonds the people and the land of Taichung closely together." If one cannot identify it sufficiently with the land, if it has never emerged in the cognitive system or if it does not readily associate with life, this p reclaims the death of a culture. As a result, the style of the city becomes blurry and indistinct, just like old Taichung City's Central District. Faced with the gradually forgotten pride and memory of the Central District's past, voice variation is generated among modern Taichung citizens' recognition towards their city's style. However, this also stimulates introspection and momentum in Taichung, in turn becoming the nutrients for Taichung's abundant creativity and characteristic.

Editor / Leain

“Taichung's creative energy is embedded in its diverse and multifarious life”

Head of the Architecture Department, Feng Chia University.
The Director of the Institute of Architecture / By Lee, Shwu-Ting

Taichung is located at a pivotal position of the north-south transportation in Taiwan. With good weather and ideal geographical environment, it cultivates many incredible features, such as parks, green boulevards, sun pastry, tapioca milk tea, and many incredible and romantic events happening everywhere. The richness of Taichung is just like a sun pastry, whose thin dough, light yellow skin, and nostalgic filling is sweet and not oily, solid and tasty. The chatters and twitters of Taichung people leave us with pages of memories, witnessing the ups and downs of the city and the people in history. Taichung has a complete and continuous historical record. The city’s image is clear, with many well-preserved historic monuments, from traditional Chinese architecture, the paths built in the Qing Dynasty, the quarters and government offices built during the Japanese rule, the aqueducts and parks built by the Japanese, the American buildings constructed in the Republican Period, the rise of the Central Picture Cultural Studio, and so on, each of which articulate the basic building blocks of Taichung, offering materials for creative design and endless inspiration. n Taichung, creativity is not a noun but a part of Taichung people’s life. The old Taichung is no longer the city in people’s impression; instead, it has transformed into Greater Taichung, a more complete and bigger city. Cultivated by diverse culture in the course of time, its potentials are eager to flourish, with no patience to wait any longer. In the city, new and old cultures interact and intermingle, constituting the Greater Taichung a city of humanity and high-tech. The potentials and determinations of the Greater Taichung outdoes Taiwan’ framework. Illuminating in the global arena, it is developing its own features.

In the book Creative Taichung, many of the designers are either the alumni or former teachers at Feng Chia University. They are all pioneers in the industry, offering unique views and diverse designs, demonstrating postmodern eclectic styles. They are enthusiastic to promote the potential of Taichung design, allowing people to know the energy of Taichung creativity and cultural innovations, the fruits of integrating creative cultural activities into our daily life.

“Paying attention to the aesthetics of architectural space is injecting creative vitality into Taichung.”

Chair, Department of Interior Design,
National Taichung University of Science and Technology / By Lan, Ju-hung

Greater Taichung is located at the center of western Taiwan, with Miaoli in the north, Nantou Mountains in the east, Taiwan Strait in the west, and the Changhua Plain in the south. With mild and comfortable weather, complete daily life amenities, and convenient transportation. Since ancient times, Taichung has been renowned as a cultural city, with prolific historical assets and aggregation of professional talents. In the recent years, historic buildings such as the Mayor’s House, Prefecture Office, Broadcasting Bureau, Butokuden of Taichu Prison, and more have been opened to the public or operated in a BOT manner after refurbishments. It reanimates not only the historic buildings but also offer people opportunities to share the cultural assets and collective memories of Taichung. Chaihsing Villa, Hsiau-Yun Villa, and the Wufeng Lin Family Mansion, which belonged to Taichung County, are annexed to Greater Taichung, adding more cultural energy to the city.

In recent years, there have been diverse cultural activities, such as the concerts by the tenor Carreras, cellist Yo-yo Ma, pop singer Lady Gaga, Vienna Philharmonic Orchestra, and more. Other constructions, such as the Taichung Cultural and Creative Industries Park, Warehouse 20, National Taiwan Museum of Fine Arts, Taichung City Seaport Art Center, Taichung City Tun District Art Center, and more are active in sponsoring diverse cultural activities and performances. Meanwhile, the Fulfillment Amphitheatre, the Green Park Outdoor Theatre at Taichung Museum of Fine Arts, and street performers’zones bring people different kinds of performances, adding spices to the city. The Taichung Metropolitan Opera House under construction will become an important base for the cultural and artistic performances.

CONTENTS

A Leader of Supreme Lalu Aesthetics

CCPLAN DESIGN CO.,LTD. / S.J. WANG

Design is about bringing back natural and comfortable core values within creativity; therefore, today's abstract and extravagant trend is tacitly concealed outside the architecture. Inspired by Flow Space, the fluid two-dimensional space flows within the building and allows the design lines to flow smoothly within the metaphoric arches in the atmosphere. Like a breeze, it blows past the forest, water surface, architecture and space to swirl and flow freely and comfortably. The creative admiration experienced by the senses reflects the design details emanating from creation.

Text: Yves

Photo provided: CCPLAN DESIGN CO.,LTD.

S.J. WANG

Department of Architecture, Feng Chia University

Now the President of CCPLAN DESIGN CO., LTD.

1997 Established CCPLAN DESIGN CO., LTD.

Rewards

2008 BEST 100 for the JCD Design Award 2008

2008 TID AWARD Commercial Space Design Award

2009 Aboard Report — ANNUAL OF SPACE DESIGN IN JAPAN

2009 Top Ten Design Force for La Vie Taiwan

1

A Residential Building Emperor's Place

The design concept of this project is inspired by "Flow Space",which stresses that recreation is an attitude of life, like the lobby, the Eastern sky, the round heaven and earth accentuate an orthodox form, while 13 lanterns in the center symbolize the propitious number of 13 in Eastern tradition. Not only isthe vocabulary representing "Year of Abundant Harvest" from the Book of Rites applied to the architecture, it also highlights the grandeur of the lobby.

1.To highlight the orthodox of forms based on the concept of roundness in heaven and earth, the designer places 13 lights as the centerpiece of the lobby, 13 being a primary lucky number in the East. 2. With Oriental concepts as the point of departure towards a spatial design, the square-shaped lounge sits in its ember color tone and commands a majestic depth, both intriguing and inviting. 3. On the wall of the lobby hangs the horizontal section of a rather antiquated tree, with intertwining patterns symbolizing the beginning of time and genesis of life.

2

3

Follow the axis of the lobby, there is a space 100m wide and 100 long, where couch areas are created on both sides, and the water scenery from the outdoor garden is introduced indoors. There is even a flame on the water, and there are two fireplaces on one wall of the lobby. A cross-section of a tree is suspended on the other wall, which conveys the symbolic meaning of Pangu creating heaven and earth. Both are Eastern concepts.

1. Bamboo-woven lanterns and cobble-paved paths aligning the green shrubbery create an elegant, seductive and welcoming ambience. 2. The favorable elevated position of the architecture and its expansive vantage viewpoint conveniently incorporate Taichung's splendid nighttime city landscape as part of the design. 3. With Flow Space as the spatial axel, water in its given container reflects a gentle flowing beauty of the architecture.

3

1

2

The ceiling and the floor corresponding to each other in their square shapes, the space in between is decorated with transparent lighting pieces to convey a sense of ease. 2. The designer utilizes the water element to soften the otherwise cold harshness of the architecture exterior, ingeniously converting it into a flow of relaxing energy. 3. With bamboo forest added as decoration, the materials reflected in the pool of water encourage an inter play of light and shadows, transforming and composing exclusively Oriental design vocabulary. 4. Stone sculpture seen in refraction of light and shadow presents itself in overlapping dimensions of real and virtual sensations.5. Lighting and spatial arrangements are brilliantly in sync with each cther to allow for the water in the pond and the plantation to soften the architectural pieces.

Pranayama Resort Reception

The grandiose opening is combined with graceful Chinese style and design inspired by minimalism and Zen. The water and natural plantation has softened the stiffness of the architecture; from day to night, the projection of vibrant colors is complemented by solemn and elegant qualities that are simply astonishing. At the same time, in order to present the building's composure and poise, wood is used in place of concrete wall structure, while the partition and French windows allow see-through perspective to connect different spaces. The transforming images and the fusion of modern and ancient ingenuities are implemented in the details to add the finishing touch o the interior space.

Integrating a certain flavor of Oriental ease to shape every corner into a natural landscape as if drawn out of poetry.

Within the definition of a luxury mansion, meticulous demand in taste and living condition is a prerequisite. However, many designs are plagued by complication and oppression caused by cumbersome decoration of the equipment, décor or arrangement. The excessive pursuit of labels, ignorance of the intrinsic quality of life and the back-to-front priorities cannot be used to create simple and abundant imageries through design, therefore designer Wang Sheng-zheng believes that in order to create a topnotch living quality, one must return to the origin. Taste is directly related to the life everyone wants to pursue; those who are passionate about nature will have a taste for nature, while those who prefer the touch and romance of French linen willexhibit similar taste in life.

1. As an extension from the architectural aesthetics of the Lalu, the space is returned to a place of beauty that is shaped and woven with minimalist lines. 2. Low-elevation or full-length windows all allow for light and shadow to readily swim around your senses. 3. To align with the surroundings via the language of nature, all conclusions are synthesized for an in-depth interpretation by way of creativity.

1

2

3

2

1. In Orient architecture, designers apply the concept of space – such as scene transposition, layering, Jiangnan garden landscaping – into the design. 2. The designer gets creative and whips up an aesthetic of character, presenting space, highlighting texture and softening nature into ambience. 3. Flowing water meanders around the architecture reflecting a maze-like pastoral yet mesmerizing landscape. 4. Carefully selected building materials are in place to strengthen the well-grounded yet mellow kind of ambience in the architecture.

Empire Twin Tower

The construct ion base is located near the bustling Museum of Fine Arts, hence the height advantage and wide perspective of the building is capitalized to infuse the resplendent night scenery of Taichung into the design. As a result, the seating area in the social lounge and the counter are all equipped with large French windows to introduce the outdoor scenery. The exterior garden also employs borderless design to allow people's visual senses to experience the tension of the design. Furthermore, simple elements such as stone and wood are applied to the architecture; through natural vocabulary and the surrounding environment, all conclusions are interpreted in depth through creativity to create the individual aesthetics that can showcase the venue, accentuate quality and blend in natural qualities.

The outdoor garden is designed as a border-free space, brimming with a charged visual force starkly perceptible to your eyes.

1

Moral Philosophy Reinforcing Design Concept

He always appreciates everything with total commitment; if the entire day of designer Wang Sheng-zheng is analyzed, almost 16 hours are spent on constant contemplation. With CCPLAN's design quality and quantity, they not only focus on the domestic market, but also want to expand to markets outside Taiwan. In order to achieve success in Asia, it must constantly evolve with time to maintain visibility in the international creative design market. Graduated from the Department of Architecture, Wang Sheng-zheng believes he is a "shark constantly on the move". The reason he compared himself to a shark is due to the design's constant explosive force, dynamism and breadth.

1.2. The beauty of harmony is finely poised between lines and functionalities by way of design, allowing the space to be the storyteller for the many splendors of life. 3.4. To embed a neoclassic style into the space construction is to frame and define uniquely creative and classically luxurious vocabulary.

2

3

4

Since he has experience in designing majestic mansions, he possesses incessant design energy that not only materializes his vision but also enables him to attain a sense of achievement. As a result, he can always surpass his designs, and his accumulated experiences drive him to break through conventions and forge ahead in other directions. As far as Wang Sheng-zheng is concerned, advancements in design require constantly reviewing previous designs. As long as one gives 100% to materializing the objectives and forge ahead with assertiveness, ones resolve and sincere faith will help to implement the essence of design. If one emphasizes too much on the timeless values of life, it will often become just a formality and one will be unable to break through the limitations of conventional style. Consequently, Wang Sheng-zheng always infuses innovation into his design and implements his unique spatial concept, such as the spiritual symbol of Flow Space. Besides rotation, axis of symmetry is also very important, because there is usually only one axis in conventional interior design. Such symmetrical form does not present any innovation or visual appeal.

1. Use simplistic elements such as stone and wood to cover the architecture is to align the vocabulary of nature to its peripheral environment. 2. Utilize natural materials for their patterns and colors and to embed them into the space to be part of the elongated lines in the space is to express an elegant and refined taste.

1

2

Therefore, symmetry is not restricted to any particular form in Wang Sheng-zheng's design; the right side will not be black in color just because the left side is black. Rather, spiritual concepts derived from space are introduced, and superficial Eastern design concepts presented by ordinary designers using lattice windows or old fashioned Chinese armchair are eliminated. In terms of Eastern architectural composition, one must first study extensively to extract its essence and apply spatial concepts such as borrowed scenery, layers and Jiannan garden etc. to the design. After graduating from the Department of Architecture of Feng Chia University, designer Wang Sheng-zheng developed a profound feeling for Taichung. After establishing his studio in Taichung, he has remained in the city ever since. For more than two decades, his explosive design flair has not only earned him distinction among the domestic and international media, he has also sculpted resplendent and creative future for Taichung's architecture.

1. Stepping into the public sphere, you will be greeted with mid-air freshness that will revitalize your senses. 2. Get hold of the transparency and openness of the building and to employ a dexterous design approach is to induct an outdoor landscape indoors to shape a sense of interior space. 3. Mellow and rustic simple colors as the baseline for the space allow the residents to enjoy an amicable interaction here.

2

3

Aesthetics is the Essence of Urban Affluence

Wang Sheng-zheng developed an interest in spatial creation at the age of five, when he and his family visited the Grand Hotel. As the doors of the elevator in the hotel opened, a spatial wall seemed to open in front of his eyes and left him with a profound impression. This was his first concept towards space, and it also enlightened his awareness towards space. The majesty, grandeur and elegance of the hotel pulsated in his heart in a deep, intimate way, in turn endowing him with inspirations for design. Thereafter, memories imprinted in his mind at the time are embodied in his numerous creations.

His profound sentiments of Taichung were accumulated during his school days and the flourishing development of architectural design in Taichung have made him decided to stay in Taichung and establish his own design studio. Although he has experienced 14 years of recession in the real estate market in Taichung, his insistence in faith has helped him to overcome countless adverse situations.

Incorporate presence, elevation and depth into the same sphere is to put taste into life and give it a flavor.

1

2

Taichung is now a major metropolis, and many international architects unveil their creations here; although this stimulates development in the city, Wang Sheng-zheng believes that talents from abroad are not necessary superior to domestic talents; rather, one must gain deeper understanding in order to appreciate their influence locally. Before, the only renowned architecture in Taichung was the Luce Memorial Chapel in Tunghai University. Now, thanks to their efforts, the recent completion of the Taichung Metropolitan Opera House and Taichung City Hall has transformed them into iconic landmarks in the city. Such international exchange is extremely important not only for the outcome, but also the process, because there is a lot to be learnt. This is particularly true in terms of the city's prosperity, because although completeness is crucial, aesthetics is a key element in enriching the city. In addition, aesthetics is also a marketing concept that binds the software and hardware tightly together; it upholds the high degree of persistence towards design and endows life with the most aesthetics value. As a result, Taichung is now able to demonstrate its dazzling visibility on the international stage after designs with cultural qualities have been introduced to the city.

1. The designer turns the charm in the elegant presence of stones into spatial posture in the space, conveying further depth and richness. 2. Space can illustrate the lines in their charged, heightened state, unveiling layers of gracefulness and carving out a character unique to the resident. 3. The walls of bedrooms are decorated with intercepting squares to further highlight the arrangements of dots, lines and planes.

3

Space is the Extension of the Philosophy of Culture

Tao Yuan-ming "No dust and confusion exist within my doors and courtyard; in the empty rooms, there is more than sufficient leisure. Too long I was held within the barred cage, now I am able to return again to nature."

As the source of creative inspiration of CCPLAN Design Co., Ltd., the overall space presents the refined essence of Wang Sheng-zheng's design; whether the goal is to be realistic or idealistic, every detail has its significance. In particular, the meaning of "boundary" has been redefined, making the transition from exterior to interior a fusion of realism and fantasy. For instance, a door is no longer a door, it transforms into a rotating wall, where the image of "opening" is used to disguise the boundary. The red cast iron partition on the side has perforations that form a sieve, allowing for the free flow of air and acts as a shade to screen out excessive light. When the sun sets, light casts long shadows in the garden through the partition. The stone wall built along the pine trees and nine small dragon stone sculptures leads to the garden and it seems to vanish into infinite space. A Tai Chi sculpture by Juming above the corner has the tip of the palm pointed towards the East, symbolizing incessant vitality.

1. Capitalizing on the environmental advantages, green and water landscapes are created in the courtyard to console the occupants' vision and spirit. 2. Perforations on the red cast iron partition facilitate the flow of air and light. 3. Before entering the room, one can follow two pine trees and nine stone dragon sculptures to appreciate the relationship between people and the environment.

1

2

3

1

2

1. The gallery's depth and length exudes tranquil and sacred qualities. 2. Each material not only emanates an image, but also a visual message.3. The Tai Chi sculpture points to the East, symbolizing thriving energy in the space. 4. The artwork collection not only enriches the quality of the space, but also enhances the mood. 5. The portrait of the Miao people exude spiritual aura in the peaceful gallery. 6. The conference room faces the verdant greenness and mirrored surface of the pond in the garden, unwittingly calming one's mood. 7. Transformation of light demonstrates vivacity in life.

3

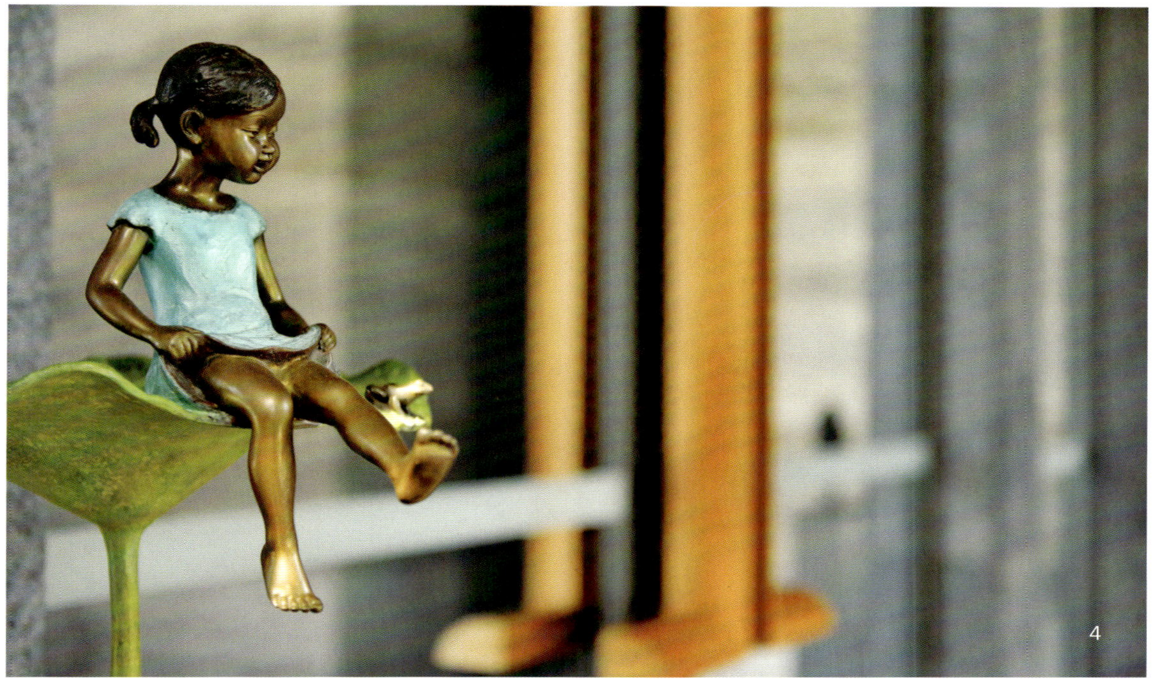
4

5

6

7

A Dreamer Creating Legends for the City

TIEN FUN INTERIOR PLANNING CO., LTD / CHANG, CHING-PING

The design drafts of various luxury mansions construction projects drawn by Ching Ping Chang have escaped from the velocity of the norm. With artistic touch and showy manner, the designer integrates fashion with the classics, demonstrating the tension of life. His works have already created certain impact in central Taiwan and the mainland. Chang's use of color is creative as well as classical, skillful blending magnificent and elegant elements into the overall planning. From the details of the design, and the use of luxurious materials of decorations and the laser logo in perfect proportions connect architecture, interior design, aesthetics of everyday life, and fashion. Employing these elements with a magic a mind, he exploits their special features thoroughly.

Text: Julia

Photo provided: TIEN FUN INTERIOR PLANNING CO., LTD.

CHANG, CHING-PING

1988 Established Taipei TIEN FUN INTERIOR PLANNING CO., LTD.

Publications

Luxury, The Dragon's DNA

Rewards

1995 Taiwan Architecture & Outstanding Enterprise Interior Design Company — Gold Award
2010 Shenzhen International Space Design Award Idea-Tops — the Best Display Space Award, the Best Office Space Award
2011 UK Andrew Martin Awards — International Interior Design Award
2011 USA International Design Awards for Commercial Space of Interior , Commercial Space of Reception, Commercial Space of Club
2011 Hong Kong Perspective Awards for Public Space Design, Clubhouse Design
2011 Shenzhen International Space Design Award Idea-Tops — the Best Commercial Space Award, the Best Residential Space Award, the Best Display Space Award
2012 TID AWARD for Single-deck Living Space
2012 UK Andrew Martin Awards — International Interior Design Award
2012 Shanghai IAI AWARD — Top Award for Furniture Space Design, Merit Award for Club Space Design, Commercial Design Award
2011 Hong Kong Perspective Awards — Top Award for Living Space
2012 Shenzhen International Space Design Award Idea-Tops - the Best Display Art
2012 Chinese Jintang Prize — Merit Award for Public Space

National Mansion The Montage Dialog of the City Context

The reception center of the National Mansion in Dalian is a newly completed project. It was designed with two concepts-"cultural encounter" and"the intertwining of actual and virtual lines," both of which constituted its specialness, values, and spatial syntax. Its spatial structure was inspired by the four stationeries in the study room – paper, ink brush, ink, and ink well, which created both abstract and concrete spaces, fully demonstrating Chinese's humanistic charm. Regarding its spatial syntax, Chang injected the self of respecting the heaven and earth, the experience of the wane and wax of the moon, and authentic characters inspired by nature in the Western spatial layout represented in Chinese classical poems and verses.

1.Cultural encounter; entrance of the National's Mansion. 2.New China's humanistic charm: entrance of the National's Mansion.

1

2

His concrete style and layout were the essences that liberated both the Eastern and Western traditions. Blending the images of the paper, ink brush, ink, and ink well, he transformed them into a special space that fully demonstrated Chinese humanistic charm. Laden with the features of different cultures, he created the all-inclusive meanings of an entity through various arrangements and combinations. With the conversational space and situations, he depicted his style and connotations, displaying Ching Ping Chang's insistence design of montage. Somebody said Le Corbusier's design
was a"machine for the living-the houses with accurate functions. If you agree with this concept, you will also agree that Ching Ping Chang's architectures are the "machine of dreams."

1. Cultural encounter: gardening design of the National's Mansion. 2. New China's humanistic charm; lobby of the National's Mansion. 3. New China's humanistic charm; model hall of the National's Mansion.

1

2

3

2

Chang transforms the unimaginable in Eastern culture, such as poetry, painting, philosophy, and moreinto his architecture's space. Breaking through the bondage of current spatial elements, Chang has defined luxurious mansions in innovative manners, leading us into imaginable and sensational realm.

3

1. A fantastic visual realm: multifunctional lobby of the National's Mansion. 2. New China's humanistic charm: hallway of the National's Mansion. 3. Hallway of the National's Mansion.

Dragon's DNA the Essence of Evolution in Traditional Culture

Ching Ping Chang mentioned, "In my architecture, I want to emphasize the introspection of a Chinese or Chinese architecture in my design. With endless practice, I attempt to express the temperaments and vigor inside me to things outside." Likewise, I pay special attention to the perspectives of seeing outside from the inside of architecture. This is something different from Western architecture, which shows off the owners' identity and social status to passersby.

1. Tree of Life; Entry Area of Dragon's DNA 2. Peony, Buddha's hand; Lobby of Dragon's DNA.

1

2

In his award winning work, Dragon's DNA, Chang summarized the idea of a"five-entrance complex." With approaches appropriate to modern Chinesebuilding. In a five-entrance complex, there was a gate in each entrance, which was connected to the gardens, pavilions, balconies, chambers, halls,and yards. In the city, especially places along the main roads to the prefectures, it was an ideal habitat for the rich and famous. In wisdom, and faith ortemperate, kindness, courteous, restrained, and magnanimous, which had been the principles of conduct for the Chinese people.paths were wide andmeandering, cultivating virtue. As a result, the people moving around were rational and learned people. Pleased with their living environments, theywere kind people.

1.Crystal peony, peal wall, Buddha's hand: Dragon's DNA. 2. Multifunctional hall in the Dragon's DNA. 3.The imitation of the poetic and artistic space in Chinese paintings: VIP lobby of the Dragon's DNA. 4. Model hall of the Dragon's DNA.

1

2

3

4

Situated in different markets, the designer still designs with the same attitude. With his insistence on his professional know-how, he accumulates his learning from culture and has it fully developed with modern means. Indeed, it is an interesting job and experience. How to preserve culture and history from disappearing in the rapid change of the environment in the milieu of the alteration of the old and the new? In addition, the designer integrates Chinese and Western cultures to create a new environment to preserve the traditional culture in architecture and rejuvenate it through architectural design. Change reforms history with innovation to preserve the old charms, bringing new lives to old things. By challenging history and the differences of Eastern and Western attitude, the designer has discovered new and different positions for Chinese architects between East and West; old and new. Blending the philosophical thinking of the East and Western lifestyles, Chang injects new meanings to the environment, allowing pedestrians and the space's life force to demonstrate innovations that are more appropriate to modern life.

1.2.3. The entrance of Tien Fun Interior Planning's office. 4. The supervisor's room, Tien Fun Interior Planning's office; Taichung City. 5. The reception area of Tien Fun Interior Planning's office.

1

2

3

4

5

Ching Ping Chang believes his design style should contain a Chinese spirit, making progress in a time of Western influence. In his designs, he eliminates the bondage of traditions and ideology intentionally. Moreover, he creates new rules for the game to express the cognition of beauty, the respect of nature, and the gratitude to life with Eastern design. In his architectural methods, Chang loves to create Chinese humor by twisting things. It seems to be easy at first glance: however, the structures are highly complicated.

The model house of Un Yan.

Finally, architecture is his ultimate means of confession, a humorous but earnest philosophy. With a mild attitude, he expresses flexible and unconventional pragmatism. Transforming the traditional norms of five-entrance, he explores the profound dialog of the spirit in the concrete space of modern times. Integrating elegance, innovations, and daring elements, he forges a stunning aesthetic space, capturing the elites' imagination immediately. Of course, his designs won high acclaim from different sectors, making each visitor relax in their senses and clean in their souls. Space and beliefs in order to pass the Oriental aesthetic task, different from the traditional concept of creative thinking on the East Totem "Simple Gorgeous naive and beautiful" design style interpretation of the new Oriental humanistic feelings.

1

Simple convergence simple and unadorned, to streamline the selection of materials for attitude, a kind of extreme warm warm connotation of humanistic adhere to the performance of every corner of the space. China and the United States is an in advertently revealed the true nature of the beauty of the East through the process carving delicate spirit to meet the modern and practical design techniques, series being in space. Between the simple grabbing and beautiful luxury, experience the life of the initial and final moved.

1. The acceptable and familiar luxury: living room of Un Yan's model house. 2. The dining room of Un Yan's model house. 3. The living room of Un Yan's model house.

Glory Tower Boundless Internationalization

What is the difference of Ching Ping Chang's design with other designers? In fact, the answer is simple-they are different in attitudes. In the project Glory Tower, we can clearly see Chang's integration of the East and the West to create a boundless international space. In the building structure with unique style, Change gives the building a smooth layout and the spaces with identifiable functions. On practical basis, the building displays the characteristics of the future world. He also creates the East visible to the Westerners; the West visible to the Chinese. Mature people tend to be nostalgic; however, the young people are eager to experience the latest fashion and accept the all-inclusive luxurious space.

1. The reception center of the Glory Tower. 2. The study room of the Glory Tower. 3. The elegant and exquisite classical charm: reception center of Heleonse. 4. The entrance of Patek Philippe. 5. The restaurant of Patek Philippe.

1

2

3

4

5

"I like classical objects and beautiful things, and even ephemeral fashionable elements. But I don't follow the latest trends everyday life,"Ching PingChang expressed. Observing the environmental change in central Taichung, Chang keeps a close eye on the developments of fashion, exhibition, commercial space, installation art, and humanistic activities. Chang said, "The changes and growth of Taichung inspire me in spatial design and stimulate my creativity.

1. Continuing the legacy and making innovations: reception hall of Wan Jin Yuan, Chengdu. 2. Continuing the legacy and making innovations: entrance and hallway of. Wan Jin Yuan, Chengdu. 3. The old soul familiar to the Eastern and Western world: reception area of Wan Jin Yuan; Chengdu. 4. The Chinese touch of ArtDeco; banquet hall of Wan Jin Yuan; Chengdu.

2

3

4

Many people said Ching Ping's design was luxurious. However when luxurious encountered interior design, how did they ignite the sparks of creativity? For people that understood Ching Ping's works, luxurious did not necessarily implied fashion but it included fashion. Therefore the terrain of his design did not limited to brand name, garment, and so on. Instead, it was a natural attitude lifestyle and the manifestation of beauty, including the atmosphere surrounded the people, who totally blended in it, including the choice of accessories, dinning room, and so on. Returning to the basic facet of design thinking, Ching Ping departed from Taiwan and presented complete interpretation for luxurious mansions for the modern Chinese.

1. Within the process of simplification and transformation, eastern vocabulary still pursues the appeals of ambiance and charisma. 2. With the right material and texture, the space will naturally radiate vivacity.

2

In Chang's hands, he has transformed luxury into the pursuit of lifestyles. With an acceptable and familiar manner, he demonstrates his diverse Chinese aesthetics values, emotions, cultural trends, and his natural born artistic soul and temperaments. His appreciation of beauty and creative interpretations signed on the things he see and the objects he touch. From unique aesthetic perspective, he integrates his innovations in his magnificent designs. Blending Eastern and Western cultures, he gets away from traditional expressive manners by presenting uncommon contents, leading new trends in the fashion.

1. With well-executed transparent and refraction techniques, one can see sceneries everywhere. 2. Light and sculpture are an extension of the design concept. 3. Designs a key that widens one's perspective in space and life. 4. The fusion and arrangement of artistic conceptions make the space more enchanting.

2

3

4

1

Bright Open Space

From communication, understanding, and the demand of details in everyday life it extends contrasting ideas of connection and extension. From the opening /concealing space, the contrast / gradation of light / darkness. It creates a flow of stacked spaces and the diffusion of corresponding spaces. Creating corresponds and contrasting relationship for these domains, it allows people to change their perspectives of space. It also leads them to transcend existing imagination to fulfill the wishes hidden in their minds.

1. In the open layout, the layers and depth are enhanced through plane structures and visual arrangement. 2. Furniture acts as the intermediary of the arrangement and continues the majestic demeanor. 3. The green wall is complemented by mirrored surfaces and light to present liveliness.

2

3

1

2

1. The unostentatious furnishing exudes warmth and emotion. 2. Stone materials are used in the bedroom to accent rugged tenderness. 3.The study features dark color tone to accentuate the room's elegance and peacefulness. 5. The bespoke environment enhances the value of the space. 4. Due to the see-through arrangement, life is more interesting with different perspectives.

3

4

5

The Pastor of a Space that Built His Dream

BAN MU TANG ENVIRONMENT INTEGRATION / HO,CHUAN-XIN

Architecture, like plants, grows up from the horizon. It has the power of piety, respect, and modesty, cultivating intimacy between the land and people. The environment affects architecture, which bridges the interrelation between people, heaven, and earth. Life weighs, showing beautiful pictures.

Text: Jeremy

Photo provided: BAN MU TANG ENVIRONMENT INTEGRATION

HO,CHUAN-XIN

Graduated from Department of Architecture, Tunghai University
Founder and Creative Director of BAN MU TANG ENVIRONMENT INTEGRATION
Partner of Jiang Wenyuan Architects Agency
Executive Vice President for Lvcheng Construction Co., Ltd.
Executive Vice President Zhijian Construction & Development Co., Ltd.
Lecturer of Department of Architecture, Feng Chia University

1

Bao-Shan Pavilion was the Beginning of Everything

Located in Baoshan, Hsinchu, Banmu Villa was planned and built for the needs of office and business space when Banmu was first established. By the entrance, there was a City God Temple covered with new square fencing. In front of it, there was a concrete guide wall with two round holes. A visitor had to walk around the guide wall in order to have a better view inside. The walk started from this side of the building. At the end of the guide wall, there was a green field grown with green trees, connection to the mountains afar. Passing by a waterfall, there was a wooden information desk at the left and a concrete wall extending outside the room. Inside, the floor was as high as the matured rice. In the room, it was about 100 m^2 in size. Built with the primary materials, such as concrete, steel, and glass, it presented the beauty of simplicity. After pushing the gate open, the villa became a pavilion, where wind and light passed through naturally, without any boundary. At the neighborhood, barefoot farmers came and sat on the platform under the camphor tree to take a rest. The egrets hopping in the field stretched their heads inside as if peeping for something. Then, everything – the building, nature, and people were connected to each other, creating an intimate relationship. Banmu had made a start in this manner.

1.The dialog between the architecture and environment. 2.The greenery . 3.Transparent, it lets in light and landscape.

2

3

International Bodhisattva Sangha the Integration of Buddhism and Architecture

The IBS Buddhist Temple was located at Dali, in a crowded city center, surpassing the relationship of the urban texture and a traditional Buddhist temple. Nature was surrounding the building, preaching "dharma" with the natural environment. In the temple, the routes were arranged with the image of "the path to Tao" narrated in a Buddhist sutra. From the first to third floor, visitors had the opportunity to encounter nature. Employing the"responsive essence" of concrete and the experience of the material, the designer constructed a tranquil atmosphere. It derived the attitude of the building as solid, symbolizing the enlightenment of dharma and inspiring the respect to Buddha. Matching the daylight with the flowing pond, daylight, trees, and a pond were introduced to construct the "habitat of the soul." It complied with the idea of "finding your heart" as taught by the monks in the temple. The building and dharma shared and tolerated everything because of chance.

1. The beauty of simplicity and leisure. 2. The symbiosis of architecture and greenery.

Dosha, the Response to the Integration of the Spirit and Art

In response to the needs of the space for relaxing the mind and body in the readjustment area, the entire structure of Dosha was projected as an introverted building. In front, there a double-layer concrete wall to respond to the noise of the readjustment area in the city, providing Dosha SPA with the function of privacy. It also left a blank between the city and the hearts. Likewise, the building had become the medium between the city and the hearts, constructing the experiences of the five senses beyond the realms. Most of the building was lit by the light spilling in through the courtyard, surrounded by the camphor and birch forests and a pond. Situating in the midst of these natural elements, the body and soul still felt comfortable in the midst of the city.

1.The correspondence of light and shadow. 2.The definition of the lines and dimensions. 3.The architectural essence that recalls nature. 4.Nature nurtures our body and soul. 5.The trace of architecture.

1

2

3

4

5

International Buddhist Association South Island New Zealand for the Buddhist Community Centre

In the religious and urban culture in the West, people "preached through art," which was employed as the idea to preach Buddhism, a sect of Eastern culture, with the temple. Using stones from New Zealand as the major building material, the Buddhist logo was carved on the façade of the main hall. The bas-relief of the Buddha was reclining at ease. He preached casually without any predetermined form. The trees, grass, and the brook had become the "natural" benign media that dissolved all the barriers. This building won the National Award for Architecture and Green Architecture in Taiwan.

1.Share the glory with heaven and earth. 2.The texture corresponds with the urban sensation. 3. Contain the authentic meanings ofsimplicity. 4. The casual arrangement of light and shadow.

4

1

2

3

4

1.The beauty of ratio. 2.The charm of light and shadow. 3.Calm and tranquility. 4.The tension of dimension.

The One Nanyuan Land of Retreat & Wellness the Idea of New Architecture

The architect maintained the principle of refurbishing the old building to preserve the original layout of Garden Nan and inject new thinking in its design. With the four seasons as the source of inspiration, it integrated with the existing landscape to demonstrate the charm special to each season. Associated with the idea of "garden picnicking", it created an aesthetics that blends into nature. The function inside the building was transformed into a humanistic inn by maintaining the original layout and structure. The one to one ratio of the bathroom and the bedroom interpreted a slow tempo ideal to an inn. It was possible to allow lodgers to"relax themselves completely"- a marvelous experience to them.

1. The bamboo and shadow conserve the essence of the East. 2. The architecture is laden with Eastern beauty. 3. The syntax of the connection of the media and environment. 4. The symbiosis of light and shadow.

1

2

3

4

1

1.The implications of the bamboo and people's tranquility. 2.Transparent keeps company with nature. 3.Pure and clear like water. 4.The dialog of sleep and environment.

Ban Mu Tang Headquarter

Originally, it was a traditional farmhouse located between an industrial town and a new town center, with paddy fields and fruits trees around. Rented from its owner, the surface of the building was refurbished and re-defined to become the headquarters of Banmu. In order to promote the idea of"work as a part life", the designer designed different spaces of everyday life with various functions to place to allow people experience the space and environment in it. As a worker opened the door, he or she would walk right into the field. Meanwhile, light and shadow had easy access to the building as well. In addition, people could also take a walk near the pond barefoot to feel the biodiversity. Along the paths between the rice fields, the architect also built many wooden plank ways, offering office workers opportunities to take walk once they got out of the office. Around the building, there were many balconies that were bigger than any of the conference rooms there. The designer also preserved the low brick walls covered with green vines and most of the fruit trees.

1. Befriend the forest; brainstorming. 2. Work befriends the country and forest. 3. Extending the boundary of natural design. 4.The symbiosis of line's ratio and environment.

1

2

3

1

In addition, he also transplanted many old trees to add more greens to the office. As a result, all employees were working and creating in the midst of a green park. Inspired by the greenery, they always came up with wonderful works and ideas that brought people closer to nature.

1. Transparency extends power. 2. The intimate dialog between water and plants. 2. The harmonious expressions of the stones and green grass. 3. The company of leisure and greens. 5. The trees stand upright at ease.

2

3

4

Je Shui Pavilion

This is the corporate guesthouse of Banmu, an important chapter on delineating environmental integration. The building was located by an old warehouse in a main route in Zhubei. Extending the experience of refurbishing Bao Shan Villa, the designer opened up the original cover and built a "pavilion" with openings on all sides. In the building, concrete walls stretched inside from the yard divided the space into the dining room and living room. The dark color tone compressed the vision and introduced more natural elements from outdoor. Inside the building, the space carried life and humanistic atmosphere. Outside, it was covered with nature and environment. The major business of Banmu was architecture; it was its corporate mission to consolidate the idea of becoming one with nature.

1. The architecture converses with the environment through transparence. 2. The materials respond to the environment's nature. 3. The authentic treatment of the ratio of light and shadow. 4. The harmony of natural light and indirect light. 5. The relationship of extension, extensiveness, and freedom.

2

1

4

5

Da-Ping-Wo Village

Located in Hsinchu, the Da Ping Wo Village was a newly developed solar term architecture. With a mild dimension suitable to people, it displayed the Ban Mu Tan team's insistence on creating a friendly space. Beginning from clearing up the environment, the team preserved the old ducts and wetlands. Moreover, it used natural construction method by laying rocks along the brook's banks. In addition to preserving the 60 old trees, they planted another 100 native trees to create a pastoral atmosphere. During the process of clearing up the trees and construction, the famers still irrigated the field and harvested rice. Wild animals, such as owls, still visited the place, provided adequate evidence of Banmu's insistence on caring and respecting nature. Situating in the mountains and enjoying the daylight, people should be thankful to the harmonious arrangement of nature. The design and idea of Banmu's environmental integration team showed not only the respect to nature but also fulfilled the attitude of architecture – the attitude of working with life in a serious and responsible manner.

1. The symbiosis of the architecture and natural environment. 2. Sunlight spills in leisurely. 3. The lines become rich with the wax and wane of light and shadow. 4. Respecting the natural ecology.

1

2

3

1

2

3

1. Retain the natural factors and the surrounding architecture is the insistence of being friendly to the environment. 2. The beauty of simplicity. 3. Dapingwo Village. 4. The architecture breathes in correspondence with daylight and the mountains.

4

Fu-Gui Sanyi Art & Culture Gallery

Situated in the misty mountains, Fu-GueiSanyi Villa is constructed as a box in the proportion of 1 to 3 in architecture and landscape, with the idea of "collection" of mountains and greens, mist and rain. The center of this box represents the nature. The huge trees, rocks, and plentiful pond water, and changeable weather creates tension, like the museum's white walls that provide ample space for the art works to present themselves. The architecture becomes the backdrop in the design of Fu-GueiSanyi, making the whole place perfect. The naturalness in Sanyi has naturally become the main character in Fu-GueiSanyi, with the building as the stage. The architecture accompanies visitors to see the spectacular rocks and serene ponds. In autumn, the summer flame tree blossom dyes the mountain city yellow. In spring, the fallen purple flowers awake the blooming Tung blossom. When the seasons shift, the building transforms nature into more magnificent landscapes.

1. The flamboyance and artificiality of manmade architecture is minimized to preserve the precious natural landscape. 2. The large span of grey precast concrete is the perfect setting for lighting demonstration.

1

2

1

2

1.2. The interior space is predominantly grey or black to reduce the presence of architecture and display the artwork. 3.4. The panoramic window allows natural light and landscape to be introduced indoors.

3

4

A Performer Containing Diverse Energy

HARMONY DESIGN ENGINEERING CO., LTD. / ZENG, WEN-HE

Taichung has a warm humid subtropical climate, with an average annual temperature of 23 °C, and average annual rainfall of just above 1,700 millimeters, it has ranked highly in "livable city" rankings in Taiwan. And Taichung is pretty much milder than other cities in Taiwan in terms of temperature. And with all the reasons above makes more and more people move to Taichung city. So the Harmony Design team decided to think outside the box and go from the local culture and spirit.

曾文和

Text: Jeremy

Photo provided: HARMONY DESIGN ENGINEERING CO., LTD.

ZENG, WEN-HE
Graduated from Fu-Hsin Trade & Arts School
1987 Established HARMONY DESIGN ENGINEERING CO., LTD.

SW Kukuan Resort Quietly Listening to the Calling from Nature and Heart

The Frangipani brings out the flavor of South East Asia leisure style. SW Kukuan is surrounded by mountains, which sketches a relaxing and romantic vacation when pampering with a hot spring bath under the Frangipani trees. The waterfall running along the wall is the theme designed for bath house. Together with romantic candles, rose petals, flora of wild orchids, and wooden wind chimes and Frangipani trees, the interior space expresses natural and romantic personality.

The glass curtain in the reception hall removes the barrier between humans and nature.

1

2

3

The ultimate purpose of dimension construction is to convert the energy of sun, air, and water into the strength and potency expressed in the interior space. Through the creativity of technology and craftsmanship, the change of seasons flows into the bath house that alters the expression of interior space dramatically. The air flows and circulates smoothly among different areas due to the openness of the interior space, which realizes the concept of environment protection and energy saving. The HARMONY DESIGN team performs perfectly on technology and art with their profound design capability and energy.

1. The transparent materials shape limitless charm. 2. Indulge oneself in the relief and relaxation with greenery reflections as well as open and transparent layouts. 3. Lights also shed the rhythm harmonized with the interior space.

The Sun Hot Spring and Resort The Charm forested by Interior and Exterior Exquisite Design

Pursuing the exquisite design concept of relaxation and inspiration, designer Wen-Ho Tseng has been out of the box of function-focused design through fusion of practicality with aesthetics. He took the approach of openness and leisure to design the premise, with the base of nature and literature spirit, starting from inner search to discovery of co-existence of nature and openness. The architecture in Taichung City has been inspired significantly by Wen-Ho Tseng's landscape designs of city, rural, forest and river. Due to the strong awareness of environmental protection, an architectural planning must cover a certain portion of greenery. The designer Wen-Ho Tseng adopted his devotion into nature as the initiatives of his design. The wide and open dimensions set minds and spirits free.

1. The open fountain landscape offers extended freedom to relax the body, and to release soul and mind. 2.3.4.5. The interior space becomes closer to the nature through the design.

2

3

4

5

The Sun Hot Spring and Resort Details are Illustrated with Leisure

The Sun and SW Kukuan in Taichung are best known and most talked about among all hot spring resorts designed by Wen-Ho Tseng. Located in the outskirts of Taichung area and occupying 14,000pings (11.4 acres), the Sun Hot Spring and Resort, with nature, fuses the charms of leisure styles of China and South East Asia. The allusions of grilles in the lobby and infinity of the waterscape immerse the building in the contemporary fashion with subtle yet impenetrable flow of Zen impression, which create a comfortable and peaceful ambience for a vacation. Black andesites give the building a subtle and gracious outlook, which support the distinctive serenity of the structure. After installation of light sand shadows, the ideas overflow along with the high ceiling for people to pause and to appreciate life.

1. The routes in the courtyard outdoor lead you to the tranquil holiday atmosphere. 2. Illuminated with the lighting, it appeals to users with the controlled but unrestrained space. 3. The smooth and warm materials and colors create the ambience for relaxation. 4. Lights are filtered through the oriental grilles resulted in poetic rays and shadows. 5. Distinctive serenity is visible everywhere in the resting rooms.

3

4

5

Boutique Hotel When Fashion Meets Modest Romantic

In terms of Boutique Motel design, designer Wen-Ho Tseng created the sense of crossing over various layers and dimensions of space. Nature is the fundamental concept of his design; forest and leisure are in his creation of a motel. He describes and illustrates the nature of the interior space freely with stories from people's hearts. Taichung Mulan Motel is the landmark building among exquisite motels in Taiwan. Wen-Ho Tseng is the first one who brings sunshine, trees, flowers and intent of water to construct a forest-like interior. Through the combination of natural atmosphere and elements, the room becomes vivid and full of the means of leisure and relaxation.

1.The details on the carved pillars and hallways create a classical charm. 2.3.4. Using the natural light and shadow, it sublimates the value of the space.

2

3

4

1. The court, designed as a simulation of the nature, corresponding the sophistication and indulgence in the interior space. 2. In the semi-circular space, the bathroom is so romantic. 3. Between the bed and the bathroom, the revolving door conceals or opens up the space. 4. A Southeast Asia style garden in the public bath brings out luxury in a paradise. 5. Curtains decorated around the bed are fine and weight less of which one may vaguely see through and feel enchanted.

1

2

3

4

5

SHUEI WU The Theme Quality Derived from the Pace with the Nature

There is no other approach when talking about design but to decide the theme, the direction of development, so as to establish the profound understandings on the object to be designed, and to review the methodology of design through the experiences of interacting with the environment. The entire course of designing is an open-minded thinking process to build a sense of time in life, and to recover the free flow of the enchantment of the interior space without any superficial grandiosity. Regardless of interior design or commercial design, the team at HARMONY DESIGN always sets an easy and leisurely tone for their work, and adds symbols of natural elements for holiday resorts located in urban cities.

1. The bold color scheme has made the business area recognizable and distinguishable. 2. The key why the interior space is so impressive is the details coordinating and integrating among themselves.3.Retro elements and modern materials echo each other. 4. Dining area is another stunning visual feast.

1

2

3

The interior spaces integrate literature spirit and the language spoken by nature. "Nature is to attract people without grandiose superficial outlook",Designer Wen-Ho Tseng said. With the international thinking and open minds, the team blend western and eastern cultures to successfully design "Nei Hu SHUEI WU" Tea and Light Food, "SW Kukuan" Hot Spring Resort,"Tempus Mandara Life Club", which are in different lines of business yet are eye-appealing and attract so many attentions due to the glamor charms and the generosity.

1. Under lights, the oriental red becomes more intriguing under. 2. The colorful birdcages randomly hanging on the ceiling decorate another type of beauty. 3. Modern materials are also capable of telling an old story. 4. The greenery flora with various heights around the room creates an atmosphere for a complete relaxation. 5. Sitting area is along the French windows, which extends the spaciousness of the layout.

1

2

3

4

5

1

2

Generous Public Area Remark the Charm of Humanity

The high ceiling and low window line extend the openness of the view, emphasizes brightness, spaciousness and generosity. The public area manifests the clear and pure characters of the building with the quality to connect with the environment and nature. The lights and breezes echo with each other. People are humbled in front of nature and able to appreciate the marks of time and seasons lef: in the space. This is so called of life, manifests the clear and pure characters of the building with the quality to connect with the environment and nature. The lights and breezes echo with each other.People are humbled in front of nature and able to appreciate the marks of time and seasons left in the space. This is so called of life.

1.Luxury harmonizes with details. 2.Lights make the generosity of the space remain cozy warm yet elegant. 3.Thespaciousnessand openness of the interior space is maintained by a paper-thin door. 4.The continuity of the structure in a large scale createsthe unbridled visual enjoyment.

3

4

1

2

3

4

The public area in each architecture and building is the symbolic of spirit and home for life. Once the concept of humanity is infused and illustrated, residents who live in urban cities may create their own life-styles freely without being confined or controlled. The interior space becomes generous and glamor when plants, waterscapes, materials, and lines are in perfect proportion and flow coordinately.

1.4. The furniture and lights with intriguing shapes and lines elevate the taste of the interior space. 2. Lights and shadows are changed randomly to emphasize the vitality of the interior space. 3. The arch structure leads to the open public area. 5. The ingenious flora is artistic.

5

Construct a Space That Has Its Spirit

Only when design combines with life may one create the value of aesthetics, elegancy, and practicality. Other than the visible icons on the surface, the combination offers intriguing sophistications. To designer Wen-Ho Tseng, design is not a directionless wander. Design should fulfill user's heart, so that the thoughts and creation may penetrate into mind and soul, further more integrate with the space. The heart is able to look beyond tangible materials and icons, and reach the level of relaxation and contentment. The spirit remains eternally in the space.

1. The lines on marbles are one kind of visual art. 2. The layers of the layout open a conversation between life and space. 3. The tone of colors, lines and proportions of each piece of furniture match with the comfort and serene interior space. 4. The right proportion and position of pieces of furniture outlining the spaciousness and openness of the interior space. 5. The combination and integration of needs in life and interior design makes stylishness meaningful.

1

2

3

4

5

1

2

3

4

5

Besides focusing on the contemporary architectural trend, Wen-Ho Tseng cares about the needs in life, enhances the functioning mechanism with aesthetics in the premise, and maneuvers the flexibility of the interior space. His design is not about serious space allocations and arrangements, but about the ambitions of utilizing the tone of nature freely to capture the reality in the joyful atmosphere, and to reveal multiple dimensions of life.

1. The function rooms are separated by glass without interrupting the continuity of the layout. 2. The continuity of the floor tiles gives the interior space openness and spaciousness. 3. The ceiling of the dining room is cover with mirror-like materials which extend the interior space vertically. 4. Shower area is covered by solitary tiles. The continuity of the layout emphasizes the spaciousness. 5. The open-style bookshelf sketches powerful lines extended horizontally.

1

2

3

4

1.2.3. Lights indirectly projected create a serenity and peace in the room. 4. The soft tone of colors and quiet materials give the interior space unearthly graciousness. 5. Life is enjoyable and sustainable with elegancy yet grandiosity . 6. The simple background leaves unbridled imagination to drive a wandering life.

6

Louis Kahn once believed: light is the language for the conversation between human and God; it is also the manifestation for both humanity and divinity. The masterpieces of art ought to be appreciated under the natural light, so that the human, art, and nature can be melded in the common space and interact with each other. Along with the exterior change, a variety of expression is generated in the interior space. The stability of life consolidated by sky, earth and nature awakes minds and souls who become responsive to the beauty of lights.

1. The French windows create irregular layering on the façade. 2. The unorthodox method creates a spacious hallway for the guests. 3.4. Lights, considered as ornaments, enrich the space effortlessly.

1

2

3

Hospitality Trend Setting Artist

JOY INTERIOR DESIGN STUDIO / CHOU, YI

With the obstinate and unruly characteristic of an artist, he searches for unique soul and life within the conventional architectural framework. Designer Joy Chou's spatial structure is infused with abundant cultural ambiance; planning is based on different spatial characteristics to accentuate the extraordinary aesthetic appeal. The high degree of experimental innovation is unconventional to the observations of ordinary designers. Through the perspective of art, the original appearance of architecture is sought for in the realm of culture, where language, culture and body are converted to transform the building into a one-of-a-kind giant installation art.

Text：Kiwi

Photo provided： JOY INTERIOR DESIGN STUD

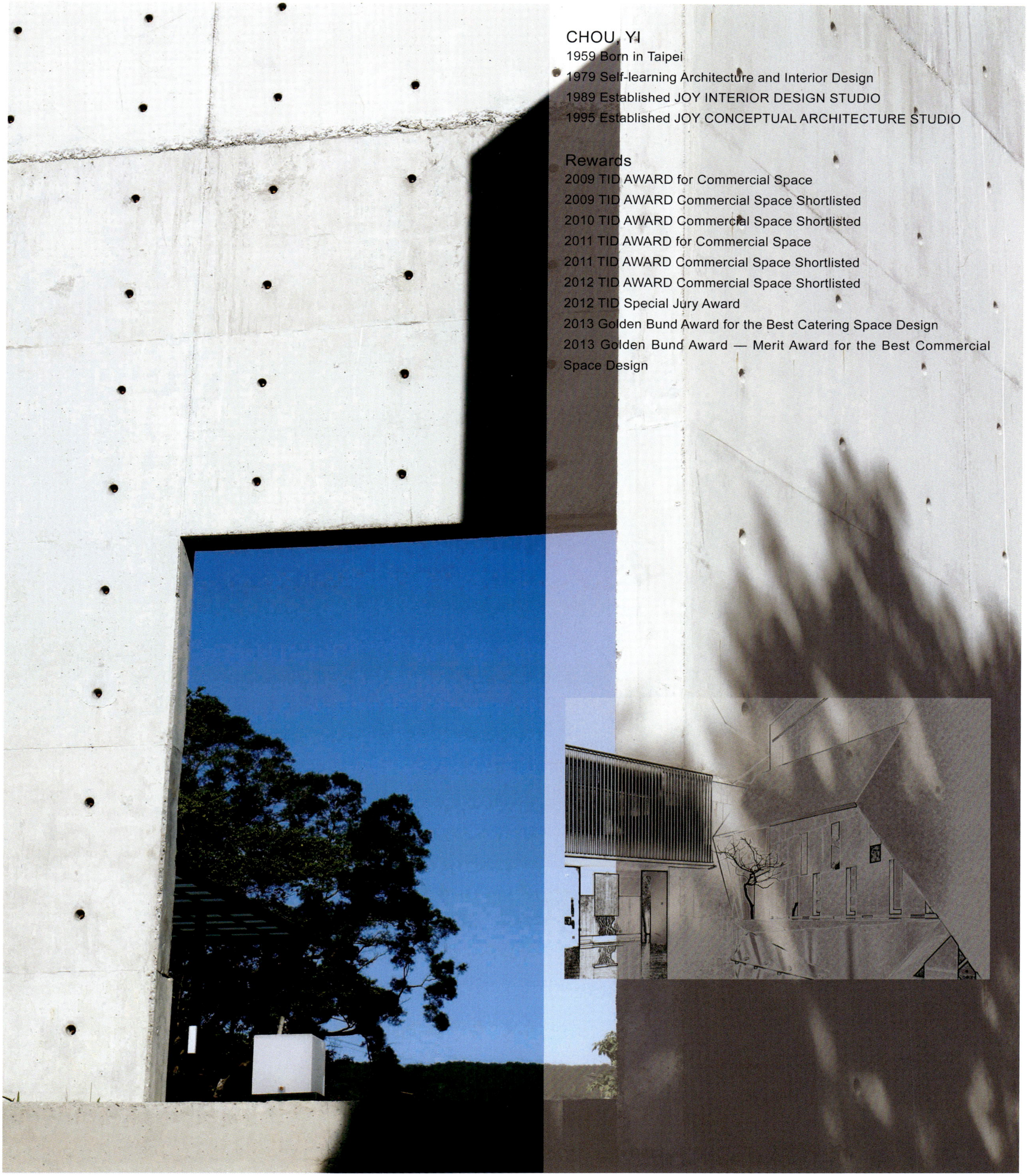

CHOU, YI

1959 Born in Taipei
1979 Self-learning Architecture and Interior Design
1989 Established JOY INTERIOR DESIGN STUDIO
1995 Established JOY CONCEPTUAL ARCHITECTURE STUDIO

Rewards

2009 TID AWARD for Commercial Space
2009 TID AWARD Commercial Space Shortlisted
2010 TID AWARD Commercial Space Shortlisted
2011 TID AWARD for Commercial Space
2011 TID AWARD Commercial Space Shortlisted
2012 TID AWARD Commercial Space Shortlisted
2012 TID Special Jury Award
2013 Golden Bund Award for the Best Catering Space Design
2013 Golden Bund Award — Merit Award for the Best Commercial Space Design

Continuation of Experimental Creativity

Nicknamed "workaholic" by his employees, Joy Chou almost never rests throughout the year, his only leisure sport is swimming; once swimming is done, he quickly rushes back for the next round of thinking and creation. Being so diligent in the field of architecture design, when asked if he sometimes also runs out of inspiration, he replies, "So far it hasn't happened yet. Perhaps after another decade, when my brain ceases to think properly, I won't be able to create work then." From his constantly surprising creative spaces, one is most curious about the source of his inspiration. According to Joy Chou "Most inspirations come from artworks, literature and social events. For instance, the 228 Teahouse and my No. 0 Apartment both challenge the possibilities of experimental concepts. Elements derived from numerous events become installations in the space to allow people to enjoy a unique spatial experience. This is a transformation and impact on the senses."

The green garden imagery at the outer perimeter of the building serves as a mood transition before entering the interior.

In Joy Chou's designs, one often finds copious application of artworks; just like his living quarters, retro, nostalgic furniture items are stacked everywhere. These came from the designer's older sister, who happened to have a batch of antique arts that she wanted him to help collect. Over time, Joy Chou also took a liking to art and began to appreciate the magnificence of these creations; hence, he applied them to commercial spaces that later become highly popular. Many art galleries and pubs in Taichung were created by him, and they in turn exert profound influences on modern Taichung's teahouses and restaurants. For instance, Chun Shui Tang on Jingming 1st St., the most popular Japanese hot pot restaurant in Taichung – Karuisawa, and the culturally characteristic Natural Ha-kka Cultural Restaurant are all his designs. They not only transformed the eating and drinking culture in Taichung, creative energy is also injected to present a unique colorful appearance.

1. A water image is retained for the building's exterior to create a sense of lightness. 2. The dark-colored stone and wooden color complement each other perfectly. 3. Trees are planted to create a bamboo-forest like experience in the space. 4. Another architectural form is created within the building to challenge conventional spatial thinking.

1

2

3

4

Precision in Architecture

With a background in art, besides commercial space design that everyone is familiar with, he often designs creative venues for art galleries or organizes exhibitions from time to time, where he breaks the convention of simply hanging all the artworks onto the wall that is illuminated by simple lighting. Instead, his unique presentation technique involves placing clay arts in clay to evoke association between the artwork and material and present them in a peaceful manner, as though the artworks are about to be born. Such unusual art exhibition endows the artworks with vivacity and energy. Whether a bustling venue suitable for family gatherings or tranquil art galleries that calms one's heart down, they are able to present an alternate spatial experience through Joy Chou's ingenuity.

The famous Karuisawa hot pot restaurant in Taichung. The building exterior is surrounded by black grilles.

1

Regarding spatial design, Joy Chou describes it as "As in creating artworks, I feel elated throughout the different stages of inspiration and creation, because a brand new design is about to emerge. From execution of construction to completion, it is a part of the mission accomplished, and the challenge is also over. After some rest, I am again ready to take on the next challenge." The constant process of creation is like absorbing nutrients, additional information can be obtained from his creativity or clients. He says with a smile "Having designed so many commercial spaces, even I can grasp the secrets to opening a restaurant."

1. Generous application of stone creates a Zen-like cultural ambiance. 2. The stream and garden landscape at the exterior create a relaxing and comfortable state. 3.4. A completely different visual space inside the dark-colored building structure exudes mysterious charisma.

2

3

4

Krauisawa Restaurant Daye Branch

Karuisawa is absolutely the most popular Japanese hot pot restaurant in Taichung. Since the beginning of construction, it has attracted countless bystanders trying to catch a glimpse of its glamorous appearance. After the restaurant is opened, constant stream of patrons are allured by the aura created by the design. Inspired by the concept of Japanese holiday destination "Karuisawa", Joy Chou's application of natural greenery, calming water, bamboo branches and leaves, stone statues and Buddha statues have helped to create the elegant Japanese Karuisawa atmosphere. The fourth Karuisawa hot pot restaurant continues the consistent style and emphasizes the application of lighting to create an ambiance and environment conducive for gourmet dining.

1. The Karuisawa Daye store in Taichung features creatively designed wooden mosaic doors. 2. The outdoor corridor portion is retained and planted with greenery that sways in the gentle breeze. 3. Stone decoration complements the Buddha statue and waterscape to form a Zen-like gracefulness.

1

2

Within each creation, Joy Chou always presents it through different key elements. He believes "I don't want to produce a great work; I want a successful case to surprise everyone. At the same time, I am not afraid to be imitated, because it will never be as delicate and tasteful as the original." During the time of creation, he has applied a systematic approach from the building's exterior, interior, style, clothing, food to menu design. The overall consistency has been implemented in his design.

1. The sophisticated and understated style creates a unique dining experience. 2.3. The stylish chandelier suspended from the ceiling enhances the spatial uniqueness. 4. The waterscape next to the sidewalk enables patrons to appreciate creative design while dining.

3

4

Hakka Restaurant Dali Branch

The designer capitalizes on the elevated ceiling and large span of French windows in the original building interior to sculpt a Suzhou garden-like landscape consisting of trickling streams, bamboo forest and wooden trails through his open-minded vision, humble philosophy and logical thinking. Carrying on the theme from the exterior to the interior, contemporary, chic colors of black and white form a strong contrast with the Eastern cultural vocabulary. The two elements are conflicting yet harmonious, introducing a brand new meaning to the pleasant dining ambiance.

2

Designer Joy Chou was born in Taichung City; from his humble beginnings to becoming an architect, he was not only self-taught, his vision towards arts enabled him to ingeniously apply arts to his architectural and interior design projects, thereby establishing his reputation in the field of commercial space design. Inspired by Zhuang Zi's philosophy of "Ultimate greatness has no exterior; ultimate smallness has no interior", he has created boundless imaginations in the space. Complemented by elements such as wind, light, water and greenness, his spatial designs exude a sense of comfort and human interactions that are favored by the media, in turn resulted in numerous media exposures.

1. Calligraphy becomes an integral part of interior design; it is complemented by the continuous candle lighting fixture to constantly create surprises and stimulate the senses. 2. In another dining space, the patrons are greeted by a vast span of green forest that in turn introduces nature to the interior. 3. The tree branch placed in the space forms beautiful shadows through the light's refraction. 4. A partition is used as the space divider, which constantly exudes the charismatic character of Eastern culture.

1

2

3

4

Hakka Restaurant Chaofu Branch

Unlike the other Natural Ha-kka Cultural Restaurant on Gongyi Road, this restaurant employs Eastern cultural sentiment as the main theme. The glamorous appearance of culture, mountain and water is infused to the large waterscape outside, where guests are led to the unique dining area by the shimmering ripples. Joy Chou utilizes black window grilles to present a sense of mysteriousness, just like wooden windows in ancient times. Walk inside, the permeating Eastern elements are immediately apparent, including things such as Buddha statues, wall surface decorated in the Three Character Classic, retro wooden door, flooring with cotton prints and the bird cage. Illuminated by LED lighting, the elements further enhance the abundant Chinese sentiment to showcase the restaurant's unique dining philosophy.

1.The large water screen on the exterior creates the elegant image of Eastern landscape. 2.Bright red cotton print motif, famous in Hakka culture is used on the floor to resonate with the restaurant's name. 3.4.In order to create the cultural visual image, besides the frequent appearance of Chinese characters, suspended birdcages are also used to enhance uniqueness.

2

3

爲人子方少時親師友習禮儀
香九齡能溫席孝於親
融四歲能讓梨弟於長
首孝弟次見聞知某數
一而十十而百百而千
三才者天地人三光者
三綱者君臣義父子親
曰春夏曰秋冬此四時
曰南北曰西東此四方
曰水火木金土此五行
曰仁義禮智信此五常

4

1

Er Bu Yuan Reataurant

Although it is a dining space, it is inspired by Japanese hot spring; hence, besides being a dining space, it is also an art creation attracting guests dining here. Joy Chou projected his impression of Japanese hot spring into his architectural space, where black-grey colored rusty iron is used as the exterior structure of the architecture to create a discreet space. On both sides are grille windows complemented by illumination to form a minimalistic Japanese style. After entering the interior, paintings of bamboo forest, Japanese lighting fixture, green masonry, driftwood, bamboo tubes suspended in the air and antique gates are used to bring back people's feeling of nostalgia. Different types of construction materials are applied to create an elegant dining space in the air and an antique entry door. Having a meal in this elegant place is like an experience you never had in your life, so worth it!

1. The restaurant's entrance is inspired by minimalistic Japanese style and features a bamboo forest as the landscape. 2. The restaurant space is transformed into a Japanese hot spring house, where a relaxing winding path leads the patrons to enjoy gourmet delicacies. 3. The ceiling of the VIP box features generous application of wooden decorations to create an unpretentious design feel.

2

3

To quote a description from Stanley Yen's book "Taichung is the paradise of design". Thanks to its vast territory, there is more market for creativity. From architecture to interior design, more precise designs are required. Besides, from Taichung, Hoy Chou has also expanded to mainland, which is considered as the paradise for designers when compared to Taichung, because there is simply so much unexplored land awaiting his creativity. Whether in Taiwan or internationally, Joy Chou's creative insight will prepare him for the next creative challenge.

1. Traditional door design complements the nearby landscape. 2. Gentle, soothing light is utilized to accompany the pleasant dining atmosphere. 3. The sidewalk next to the stairs also features a waterscape. 4. Bamboo forest sceneries are presented on the wall surfaces of the dining area via Chinese painting.

3

4

Nabe An Restaurant

Situated in modern Taichung City, one always wants to choose a simple, laid-back place for dining after a hard day at work. Joy Chou has designed the Japanese hot pot and grill restaurant “Nabe an” uses simple Japanese architecture to bring about a sense of spiritual cleansing for people. With the main theme of “cleansing and praying for blessing”, this is unlike ordinary hot pot restaurants; rather, its delicate, composed Japanese elegance creates a comfortable dining experience, as though one is undergoing the process of cleansing to obtain peaceful enjoyment..

1. The square exterior of the architecture is free from ostentatious decoration to present a succinct, minimalistic appearance. 2. The designer also displayed his favorite Buddha statue in the restaurant to create a peaceful, comfortable ambiance. 3. The Nabe an restaurant in Taichung also demonstrates the designer’s creativity, where lantern-like lighting fixtures are installed to generate an alternative design trend.

1

2

1

Time Congealed

This catering space is a project just completed by the Joy Chou Design Team. Intending to inherit the family business, and promote and develop noodles cuisine, Joy Chou Design endeavored to give traditional noodles the refinement and profoundness of Kaiseki cuisine. Therefore, the major appeal lies on the environment and atmosphere of the space. In a residence built in the Showa Era, it gives people nostalgic and warm feelings, represented with controlled and magnificent architectural syntax. The appearance of the building is constituted with ultra-simple lines. On the outer walls, molded iron and eroded black wooden plates are placed to shape unyielding durability, decorated with a pampered stability and temperature. On the greyish-black night view structure, the scenery and the environmental lighting paint the dark with dots and misty lights, creating the effects and momentum of diffusion on a water-ink painting. The rough texture of the natural materials gives people infinite and profound imaginative space. It leads customers into an artificial, leisure, and sensual realm that they have never been to.

1. The black façade offers alluring staggered layers. 2. The simple elements are complemented by light to saturate the space with a sense of warmth and hospitality.

2

1

2

1.2. Complemented by direct and projected illumination, the dim dining space appears more tranquil and elegant. 3.4. All wooden materials are dyed black to enrich their textural expression. 5. The suspended cast iron piece expands the visual demeanor of the vertical dimension.

3

4

5

A Magician into Experiments of Originality

HO DESIGN INTERNATIONAL CO.,LTD. / LEO CHANG

Bordeaux and Burgundy are well-known wine production regions. Bordeaux wine is produced using a mixture of grape types to create a strong fragrance, while Burgundy is relatively mellower and smoother, and both are favored by different wine enthusiasts. Born in Taichung and having studied in England, he comes from the bloodline of minority ethnic group in Yunnan and has a keen sense of balance towards art and creativity. Having studied panel beating in occupational school and industrial design in university, he did not receive regular education in interior design and thus constantly experiences setbacks. As a result, his vision is very unique compared to others. He pursues originality, diversity and constant learning; within the market mechanism of conformity, he decides to challenge the unknown and experimental boundary of design and attempts to materialize the diversity of designs through individual differences, just like Bordeaux's impressive and surprising vitality. To quote Mr. Huang Shi-jia from the Nordic Lifestyle "One cannot live without creativity." he upholds the persistent philosophy of "Design comes after idea" to introduce creativity into commercial marketing and create home living art. Furthermore, he has propagated such passion to the discipline of architecture and planning. Being born in the 1970s, LEO CHANG has chosen a completely different design thinking to expand the vision and opportunities of the design industry.

Text: Han

Photo provided: HO DESIGN INTERNATIONAL CO.,LTD.

LEO CHANG

Kent Institute of Art & Design U.K Design MA

Instructor: Chienkuo Technology University

Instructor: Tunghai University

Instructor: National Taichung University of Science and Technology

Rewards

2005 Taiwan Architecture Golden Lion Award

2007 APIDA Silver Award for Commercial Space

2007 APIDA Shortlisted for Commercial Space Design (G-MOTEL, STEAK AKAON)

2008 Taiwan Architecture Golden Lion Award

2011 TID AWARD Shortlisted for Living Space of Single-deck

2012 TID AWARD Shortlisted for Furniture Space

Unconventional Motel in Taiwan

The Stay Real Dreams exhibition held at the Huashan Cultural and Creative Industry Center combines creations from five young artists such as Mayday's Ashin, Nie Yong-zhen, Wang Jian-yang, Bu Er-liang and Gu You-wen. The contents span across music, design, photography, illustration and fashion, where Taiwanese creativity is reinterpreted through different angles. In particular, emerging photographer Wang Jian-yang, recommended personally by Murakami Takashi, exhibited the “Home Collection” to explore the relationship between body and space. The collection has won numerous European and American photography awards, while the other collection “Private Space” was photographed at the motel designed by designer LEO CHANG. LEO CHANG upholds an interesting philosophy and approach towards motel design to endow visitors with different themes and infinite imagination, as though they have been teleported to a different space and time. His designs have evoked the interest of National TsingHua University graduate students; some have even written dissertations on this topic and referred to LEO CHANG's motel design as the turning point of motels in Taiwan. First is the G-Motel in Taoyuan, where its modest design successfully pioneered the trend in the hotel design industry. “A Room a World” and “Cosplay” have become extremely popular, and the motel is now an alternative location for wedding photos.

1

2

1. Chang Shun Yun designed the Sato Motel located in Taipei City as well as its exterior landscape and lighting. The Motel exterior is lit up with as pecial lighting design, and at the rooftop one can see realistic fire effect. 2. The poster of "Dreams of Stay Real," an exhibition held in HuaShan Culture Park. 3.4. Wang Chien-yang, a photographer, created a series of works "Private Space" at Sato-Castle Hotel. Photos exhibited at HuaShan Culture Park.

3

4

1

2

3

The unique castle structure towers in the streets of Taipei to create a feeling of out of space and time and boundless imagination. It has received wide popularity among the media. Many music videos and idol dramas have been filmed here, including Da Mouth as well as Wang Xin-ping's MV, Copenhagen's Fairy Tales, Jay Chou's MV and My Queen. Online game launches, movie press conference and pop magazines like to utilize Sato Castle as the venue, in turn, demonstrating LEO CHANG's spatial charisma and his design team. In addition, besides interviewing designer LEO CHANG, reports have also been made by tourism magazines such as renowned Hongkong Walker in several issues. As a result, this has become an essential itinerary for many tourists from Singapore, Malaysia, Hong Kong, Macau and China. Sato Castle is meticulously designed, just as a director conveys a scene he has envisioned. Bit by bit the project was completed painstakingly by the team. The scenery of each room is designed and supervised by LEO CHANG, while the team he collaborated with is the professional stage prop team from Seediq Bale & Life Pi. The level of dedication that goes into his designs is clearly visible.

4

1. The Cinderella theme room has many fans and is only available via booking one month in advance. Since the opening of the Motel, there have been over 100couples that carried out their proposal or wedding photo shoot here. This is a perfect site for a dream come true. 2.3. The theme rooms -"Speed and Bliss" and "Troy" at G-MOTEL. 4. Farewell Concubine theme room reminds one of a well-known legendary history tale, and a game with the intelligent-controlled chessboard on the ceiling with movable chess pieces will reveal how XianYu, the main character in the story, goes toward his own downfall. 5. The Pirates of the Caribbean that has taken the world by storm is also a theme for rooms, where one can shift the dagger-shaped remote control at the bed headpiece and the bed, designed as the deck on a boat, will start rocking.

5

Revolution of Affordable, Luxury Consumption

LEO CHANG specializes in narrative spatial design, and his most iconic project is Akaoni (Gongyi Store and Chongde Store). This is a Japanese steak store concept incorporating the symbol of ghost into its design, where innovative, unconventional techniques are employed to create impactful, majestic spaces. In Japanese, "ghost" can be interpreted as praises of wisdom and bravery. In Japan, many warriors were given ghost titles.

1. Steak is often considered a western dish. Japanese-style steak gives emphasis to its flame roasted cooking method, which is a brand new concept in Taiwan. In a single-month, 40,000 plates of steak are served in one single store and testify to its place in the market. 2. Charging past from across the street, one cannot help but notice and ask about this impressive steak house by a designer that employs audacious design elements and color combinations.

1

2

1

For the exterior, irregularly inclined black structure emerges from the ground with dynamic force, as though breaking the restraint of shackles with ghostly force. The overall presentation attracts attention via dramatic approach. The fusion of contrasting black and red colors represents the imagery of grilling, which injects the space with energy and reinterprets the new form of affordable steakhouses. Abandoning the stereotypical Japanese motifs of precast concrete, rusty iron and bamboo, color symbols are utilized and the colors of red and black are applied onto the irregular structure to generate linkages in people's impressions. This not only helps the proprietor to develop a more efficient marketing method, but also overturns spatial design thinking of affordable restaurants, thereby creating a brand new path to affordable gourmet food.

1.The word, Ghost, is written into the name of the shop, corresponding to the "Ghost" in Blue Way Jeans, both sharing the innovative yet rebellious image and character of ghost-like master skill in the irrespective line of business. The Japanese style no longer merely evokes a decoding of bamboos, rusty iron and exposed concrete construction, but faithfully expresses the common masses' higher expectation for a mid-ranged pricing. 2. Treat steak like a Japanese warrior, stubborn in his fight against the scorching flame. The exterior black-and-red armor symbolizes the warrior charged with force; the red hot irregular light patches are the flames enveloping the warrior's body; a dramatic approach is in place to impact the street and turn heads. 3. The vocabulary of irregularity at the exterior design is extended into the dining area, with the wood patterns, gold color, and grey color tones to convey an alternative Japanese dining ambience. 4. The layered, overlapping design at the ceiling is akin to the cracks drawn by the swordfights amongst warriors, while fulfilling the function of ventilation and smoke emission.

2

3

4

Another proprietary brand "Fever" is also affordable clothing line that has accumulated sufficient popularity to become a department store brand. Fever focuses on consumers preferring alternative tailoring; the overall design consist of white space and spatial vocabulary inspired by a hospital to highlight Fever's alternative design and irregular black succinctness. Such installation art technique is rare in a department store. The fitting room is equipped with consulting room like curtains, a hospital bed and surgical lamp; functional aesthetics is transformed into surrealistic style, thereby achieving the objective of leaving an impression and reinforcing brand image.

1. Turning space into a piece of alternative installation art, Chang Shun Yun goes by his trademark independent individualistic perspectives and gives renewed interpretation of spatial aesthetics. 2. The intricate drip and its shape outline one fantastical scenario after another that is only native to dream lands, constituting a surrealistic fantasy space. 3. Different concepts are rearranged and combined to kick-start an era of creativity, in which business is no longer merely about business and space, is not statically confined to space. 4. The illumination on the surgery table will transform the vision into a twilight zone, which is slipped into strips of multi-layered stream-of consciousness, turning the innovatively tailored concepts, causally strewn and interwoven, into graceful interludes, impressive and lingering.

1

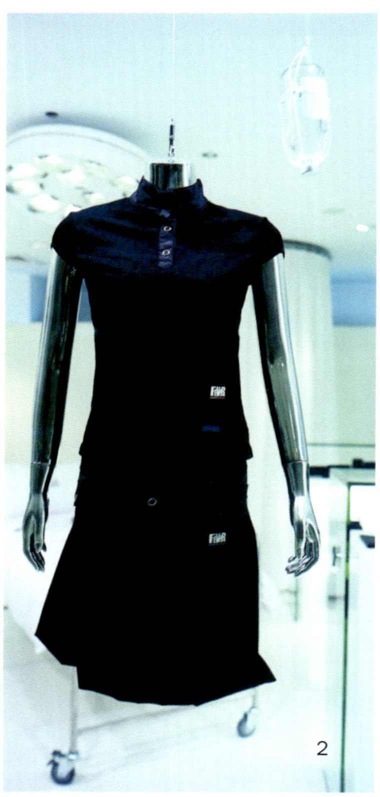

2

3

Unique Bespoke Residence

Returning to the basics of home living space, the multitalented LEO CHANG is just as capable, and designs each home to the exclusive specification of its owner; no designs are repeated. The mansion from idol dramas such as Sunshine Angel and The Hospital, as well as the shooting locations in Angel Lover, My Lucky Star and My Queen are all designed by LEO CHANG, proving his unique visual charisma in residential space design. The cases introduced here have different styles, demonstrating Mr. Chang's versatility in designing different homes for different owners.

Sunshine, water & pools, palm trees and inter-layering pure-white stylish villas are self-contained little residential universe of its own, as if transposing one to an alien land.
The entire architecture, landscape and spatial planning are all completed by the designer, Zhang ShunYuan.

1

2

3

Chang's Mansion in Beitun, Taichung is described as the most beautiful house in Taichung. The project's interior, landscape and architectural design were carried out by LEO CHANG, who customized the family's unique living and leisure space. The owner of the house is simple, straight forward, uncomplicated, confident and gentle, just like the white color of the house. Situated within a residential area in the city and surrounded by nearby buildings, the architecture is like a pearl in the desert.

1. Sunshine, water & pools, palm trees are a self-contained leisurely space for holiday makers.2. Seen from the villa exterior walls, the building is stylish in its pure-white inter-layering design.3. Tailormadefull-length airtight windows from Germany blur the interior/exterior divide. 4.5. The hallway and the dining area are separated with the designer-made curved fish tank specially created by the designer to give a pronounced emphasis on the non-existence of right angles.

1

2

3

4

A true mansion needs to be complementary to the occupants' lifestyle, intrinsic richness is more important than extrinsic richness. Although extremely busy, the successful young proprietor perceives great importance in family life and recreation. He loves surfing and RC helicopters, and the vast garden and elegantly shaped swimming pool serve as venues for him to practice RC helicopters and surfing maneuvers. The poolside gazebo features two swings for his wife and two daughters. Watching them play on the swings is pure happiness for him. The building exterior is staggered, leisurely and majestic. The pure white color continues to the interior through the large French windows in perfect harmony.

1. The building at night has an alternative relaxing appeal. 2. The night view by the pool and the pagoda. 3. Seen from the crescent-shaped balcony outside the master bedroom, the garden and the outdoor lighting blend into an enchanting view. 4. The poolside at night has an inexplicably serene ambience. 5. Together with the curves, the special ceiling and the television wall in the master bedroom, the frameless full length glass invites the distant landscape of mountains into the interior space.

5

Located in the suburbs of Taichung is a company club owned by Ming Yang Machinery Co Ltd, for the reception of foreign clients. A two-storey building nestling against the cliff, the club connects with a narrow strip of land that is the base for the company factory. Constructed around the terrain, the shape of the building is a polygon, covered amidst lush greens. The exterior stone materials, gentle and blended into the nature in their earth tone, remind one of a hermit secluded in the woods, self-effacing and reticent.

1. The structure and colors of the architectures blend with nature in a simple and low-key approach. 2.3. The architecture runs along the sites with many openings to all directions.

2

3

1

2

The large windows welcome both breezes and a view of green foliage, as if inviting a dialogue between nature and the soul. The interior TianJing opens up all possibilities for an imaginary hide-and-seek with the clouds and a grand unhindered view of the stars. The smooth and gentle timber, the rustic stones, the exposed concrete ceiling, and the open space all work together to convey a comfortable and leisurely atmosphere, what with the soothing sunshine, the sound of falling rain, the call of the wind, the chirping birds, and the green mountains against a dazzling blue sky. Follow the alleyway, get away from the meddling crowds, trace the track of four seasons and savor all the subtlety in the temperature changes in a day and in a season. Chang ShunYun, as a designer, blends space design into nature and brings nature into the space, enticing residents to leave all matters behind and completely unwind and relax.

1.2. The logs match the ceilings made with "fair-faced concrete", and the minimal use of materials and lines indoor makes the natural landscape around the house the leading role in everyday life. 3.4. The French windows in the dining room and kitchen invite the outdoor landscape inside.

3

4

1. The glass courtyard in the middle captures the stars at night creating infinite fantasy. 2. Fresh and comfortable guestroom atmosphere. 3. The open bathrooms in the guestrooms invite users to be intimate with nature. 4. Tranquil indoor space at night.

2

3

4

The archaic style of Xu's Mansion in Hsinchu emanates nostalgia and trendiness at the same time. The owner is obsessed with classicism, but he has shied away from simply replicating the classics and instead infused his personal preferences to create the chic home ambiance. The fusion of classical elements and contemporary techniques has redefined the impression of classicism; the space is transformed into a giant palette, allowing family members to present their favorite colors.

1.The design of a luxurious house can give rise to variations due to the user backgrounds, experience and personal likes/dislikes. Chang Shun Yun, with his propensity for representation of diverse styles, gives a fashion approach to the interpretation of classics, and his bold and simplistic use of colors begets a trendy lounge atmosphere. 2. By way of the black-n-white rhombus-shaped stone collage as the floor, the Swarovski chandelier, the ceiling coated with silver foil, together with Baroque furniture rich in a vocabulary of luxury in contrast with the black and white, an alternative interpretation thus unfolds for the imagery of luxurious flamboyance. 3. The laser-cut armrest with no discontinued lines or curves has an emphasis on craft aesthetics, serving as an interface between the first floor and the second. 4. The perpendicular and disorderly mirrors in the stairwell extend and diffuse the extended space to create an actual and virtual feel.

2

3

4

1. Different from the downstairs black-and-white grandiose, the bold color combination of green and white sends off an alternatively refreshing classic feel. 2. Classic romantic lounge area for the master bedroom. 3. The special tailor-made furniture adopts a combination of peach and white color tone, revealing a chic yet leisurely atmosphere, different from regular classic designs. 4. The classical and elegant bathroom in the owner's bedroom.

1

2

3

4

The owners of the Guo Mansion in Taipei are a couple; besides the guestroom and study, all the spaces are interconnected and see-through. One is greeted by open scenery at every corner of the main space. Through the window, the planes taking off and landing become continuous video frames from the living room to the bedroom and vice versa. From LEO CHANG's creations, it is evident that his designs are not restricted to any particular type of style, because every space becomes a meticulously crafted art space, where non mainstream approaches directly or indirectly express the occupants' lifestyle. Consequently, the project is not only full of imagination and stories, different occupants' spaces also exude their emotions and individualist colors.

1.2.3.While taking in a magnificent view and sitting across from the airport and Taipei 101 with a river in between.

2

3

1

2

1. From all corners of the lounge space, easy access to a vast expanse of view is guaranteed. 2.3.4. Look outside the window and one can observe the landing and take-off of an airplane as if watching frames of a film. It is as if having a plane take off in the living room then fly past the bedroom or having the plane land outside the bedroom and touch down outside the living room. 5. Lying in the bathtub, one can see Taipei 101.

5

3

4

Reception Center – Communicating Product Message

A reception center is a unique commercial product used to promote a construction project. LEO CHANG draws inspiration from the marketing concepts of different projects; he abandons superficial décor techniques and employs installation art and styling vocabulary to endow the reception center with a brand new meaning. The International Center reception center exuberates rich financial ambiance; the ostentatious, iconic white diamond structure at the intersection offers avant-garde architectural appearance and implies the subconscious meaning of affluence. The gentle, undulating design extends to the interior to match the style of the furniture and partition, thereby unassumingly conveys the demeanor of a financier mapping out strategies in response to fluctuations in the stock market. The design is well executed without losing its friendliness, while the use of black, grey and white colors exudes a sense of understated fashion.

The white and horizontal design of the reception room extends the design with diamond cut surfaces.

1

1. With lines and planes, it creates a clear-cut structure to cultivate business atmosphere and context .2.The routes are designed in accordance with turns and corners, for holding dialogs and connecting with the façade and form. 3. The furniture is custom-made to comply with the variations of the structure and form. 4.The lighting atmosphere is designed specifically to match the materials and forms, constituting diverse changes in layering.

2

3

4

1

Regarding Taichung's inspirations on designer LEO CHANG, he suggests "Taichung is an important design hub in Taiwan. My colleagues and I are constantly trying to maintain the integrity of design; however, it can be continued and expanded to my field of work."

1. The project is located by the YangCheng Lake in SuZhou, best known for its Chinese mitten crabs. The hotel is part of the Stage-II construction, including 174 guest rooms in a four-star hotel and 72 rooms in a motel, plus five detached drive-in villas. Our company is in charge of the planning for the architecture, the landscaping and the interior. 2. Bird's-eye view of "Lijiang Automobile Ecology and Cultural Industry Park" in Yunnan. 3. Bird's-eye view of the hot spring villa zone in Yuxi City, Huaning County, Yunnan.

2

3

Currently, his work spans across nearby Southeast Asia and China, while his designs transcend motels, real estates, theme parks and the cultural creative industry. Focused on Taichung's design team, he is able to emanate the energy accumulated locally to enhance the value of design.

1.2. The picture above is the project in the Beixin area in Shenyang City. Located by the PuHe River. There are 32 motels, 3.detached villas, and a themed lobby. The picture below is the Shilin Motel inYunnan. When it is completed, it will be the first Taiwanese style motel in northeastern and southwestern China. Ho Design International Co., Ltd is responsible for the landscape and interior design. 3. Located in Dehong Zhou, the border between Yunnan and Myanmar, the Peacock Building blends the style of Art Deco with the localities to create an exotic feel. 4. Business complex in Kunming City, Yunnan. 5.Residential and business complex building is located in Xishuangbanna, a renowned tropical resort in Yunnan.

1

2

3

4

5

A Thinker Deep in Cultural Cultivation

Z-WORK DESIGN ENGINEERING CO., LTD. / ALEX, CHENG

"The peaceful mind purifies and cleanses the heart. The clean and calm heart can see lotus blossoms everywhere. Each lotus flower represents one part of a clean heart. Each part of the clean heart is a solemn Buddha." Alex Cheng sees "design" a process of self learning and helping others to learn. With his modest and collaborative attitude, Alex's design inspires people to pursue a pure and clean mind and self awareness of reflections in the practice of living life.

Text: Jeremy

Photo provided: Z-WORK DESIGN ENGINEERING CO., LTD.

ALEX, CHENG

Chinese Culture University Fine Arts Department
Institute of Architecture, Institute of Fine Arts, Tunghai University
Lecturer of Architecture Department, Feng Chia University
Lecturer of Architecture Department, Cheng Kung University
Urban Design Committee Member of Tainan municipal government
Chief Designer of Z-WORK DESIGN ENGINEERING CO., LTD.
Art Director of Z-WORK Space Art Co., Ltd.
Art Architect of Ming Architecture Art Engineering Co., Ltd.
Environmental Designer of C+C Union Architecture Plan Co., Ltd.

Rewards

1991 PACE Italy TECN-HOTEL Design Competition Award
1996 Interior Design Gold Award "GLASS-HOUSE "
1998 SEVEN VISIONS ELEVEN VIEWS ONE VEERED Eslite Bookstore Design Show
1998 Interior Design Gold Award
1999 Interior Design Gold Award
2000 2nd Far-east Architecture Design Award
2001 1st Price of Plan Competition for Shanghai TONY WEAR Operating Centre
2003 JCD Special Jury Award
2008 Taichung Living Space Award "TEA WORK"
2008 Award for the 3rd Kaohsiung Urban Landscape Design Selection
2008 TID AWARD for Public Space Design, Commercial Space Design, Temporary Building Design (Gold)
2009 TID AWARD for Living Space Design — Duplex (Gold)
2010 APIDA Gold Award
2010 TDC Golden Pin Design Award
2011 TID AWARD for Living Space Design — Single-deck
2012 TID AWARD for Living Space Design — Duplex (Gold)
2013 AD China Top 100 Beautiful Architecture

Profound Cultural Refinement Become the Foundation of Design

Alex Cheng's design is based on his profound refinement on philosophy and culture. Other than his self-awareness and sense of responsibility, the influence of this father, who was born in a distinguished family of An-Hui and who graduated from the Chinese Literature Department of Nanjing University, played a crucial role. When Alex was little, his father gave him a hardback copy of Song poetry. The subtle yet detail rhythm taught little Alex Chinese culture. His father also requested him to learn calligraphy in semi-cursive script style and to appreciate the beauty of the flow of each calligraphic stroke. Chinese art gradually implanted and became essential components in Alex's life, which built up the foundation of thoughtfulness for Alex's design style.

1.2.3.4. No man no design; no nature no design. Most of Alex Cheng's work portrays the fusion of sky, earth and human in artistic manners.

1

2

3

4

To date, Alex still uses a calligraphy brush to sketch in addition to pencil. All he wants is to be more personal and spiritual through the natural stroke without any technological and technical rigidity. He also expected himself to subtly link his knowledge in culture, to re-digest the philosophy of Confucius, so that his interior design may help people be open-minded through vertical and horizontal lines and several dimensions.

1. The variety of interior space is manifest when sunshine and air are on board. 2.3. The usage of rough material is the continuation of the spirit of nature.

1

2

3

A Symphony of Physical and Mind Unveils the Secret of Interior Space

Alex Cheng was born in Dadu Mountain in Taichung. His family moved several times along with his father's career change. Alex studied in two kindergartens, four elementary schools and two junior high schools. After the kick-off of his career, Alex's life gradually moved away from the urban center. He moved back to Taichung from Taipei city, and finally settled in SanYi. The experience of frequent moving prepared him to be observant and sensitive to his surroundings.

1. The lines and patterns on the materials themselves are the most intriguing decorations. 2. The greenery in the interior space is the view, the affection, and the heart. 3.4. The integration of materials and lights leads the way to the divinity. 5. No decoration at all, the natural and rough cross-section of the wash basin is one kind of design.

1

2

3

4

5

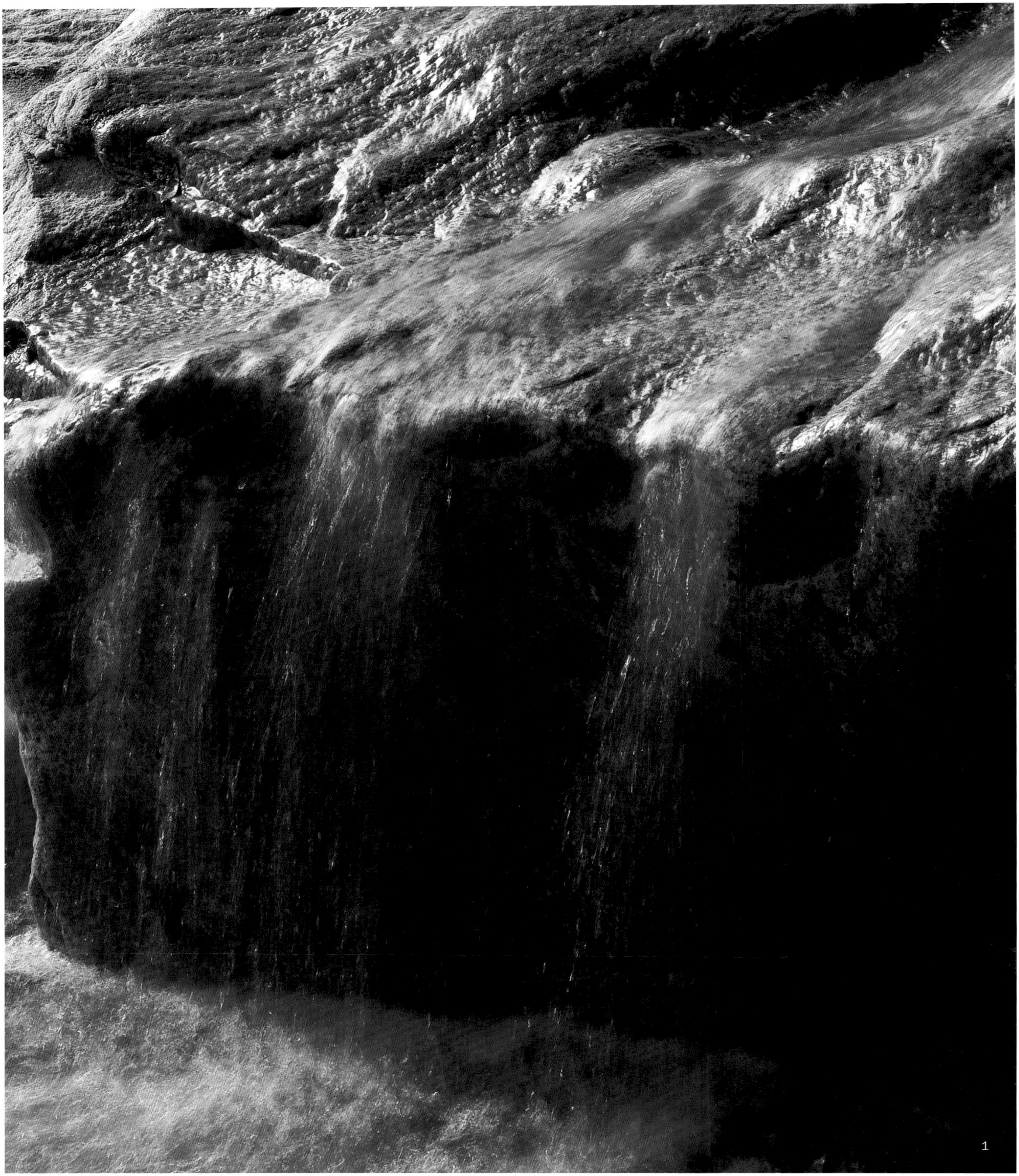

1

Alex had realized a long time ago that his experience is a life-time track record of pursuing fresh air, light, sheds, sunshine, mountain and water. He would like if the scene of nature, such as greenery, sunshine, breezes, even several of sounds and seasonal changes, could exist in daily life, as if sunrise and sunset can be seen every day. Therefore, his work, often the time, touched people by low-profile design yet full of charm. Time made his work reveal his heart and personality.

1.2.3.4.5.6. Alex Cheng expresses the charms of nature by using natural materials in the interior space. Lines are simple but visually interesting.

1

Looking at the interior space from the point of view of the natural environment, Alex Cheng believes that, prior to beginning a new design; he should empty his mind and soul till he reaches the status where his body and mind are in concordance. The pure and clear mind therefore is able to create a design that is inclusive and tolerated to deliver a deep and philosophical insight. While the residents and interior space integrate completely, the positive concept of value "For the universe, for the people, for ancient glory, for future generations" can be established.

1.2.3.4. Through various materials and shapes of the doors and windows, the change of lights and shades make the interior space spectacular.

2

3

4

2

3

Without Self, Without Image Free the Mind and Soul

Alex Chenge feels blessed for his education in childhood that had deeply influenced by the lengthy sophisticated Chinese culture enable him to form a distinctive design philosophy that is rooted in traditional knowledge and sprouts in new directions. Inspired by western style, the style of fashionable symbol collage became a trend. An interior space with subtle yet elegant charms is getting rare. Particularly, it is the needs of real estate that lead the design industry; therefore, the buildings and architecture are pre-defined and can't match the surroundings. Alex Chenge feels anxious about the current trend in the design industry. The frustration pushes him to introduce Chinese culture in his work, which he strongly believes is the key to promote the interior design industry to a new level.

1. Unpolished stone materials cover the space to emphasize the vibration of life. 2. The adequate lights create the serenity. The lights also elevate the lines on the materials. 3. The interior spaces are connected as a whole to generate serenity and spirituality.

1

2

1. The enchanting lines of the bookshelf are stylish yet classic.2. The tone of deep green in the interior space is meant to echo with the outdoor greenery. 3.4.5.6. The spacious shower area covered by stone materials. The projection of lights shows luxury and royalty.

3

4

5

6

2

Oriental Tales Infuse Sophistication of Literature into Interior Space

Alex Cheng has been trying very hard to convert the enchantment of Song poems and the melody of semi-cursive script into the charms of interior space. If you carefully examine his work, you will find that the oriental philosophy and minds are more or less embedded in each line and each dimension. For instance, the inspiration of balcony rails on the second floor of his own residence is from the poem of Song "Tune of Doleful Beauty" by Yu Li: The wind from the east blew the attic again last night, and the bright moonlight reminds me of the country I lost. The rails engraved sophisticatedly and the bricks made of jade should still be there, but the pretty face is no longer to be seen". This illustrates that it is not necessary to distant a common life from romantic poems and charms. With the infusion of oriental literature inspiration, the charms and the gracefulness of the interior space can be revealed, and develop a limitless scenery. Design, is an interpreter of languages, a messenger of passion, and a demonstration of powerful life. When the mind changes, the results are changed. Zen interior space also originated from oriental culture. Alex Cheng, however, believes that Zen - "All images are illusions", is like the free mind of Taoism and liberal behavior of Yuan-Ming Tao. Architectural materials interpret and express true Zen.

1. The water pool garden is surrounded by the buildings. 2. The vertical side of the fireplace adheres with a bamboo array collage corresponding to the Chinese style furniture in the room, gracious and clean.

Should one decide to recover the interior space where is full of the quality of oriental culture, one should be able to reach the status of no shapes, no images, free of in and out, which corresponds to the Buddha dharma "All doctrines are for purifying all hearts. If no hearts need to be purified, no doctrines are needed." This means, sometimes, there is no need to overdesign or not design at all, so that the quality of interior space itself may be recovered and harmonized with the nature, which will assist of unite physical, mind and soul. Alex Cheng's life is normal and down to the earth, just like his work. It is not difficult to feel the freedom of heart and soul through each line, dimension, and layout in his interior design. He is humble and respects life, the environment, and the universe. He is calm and modest, and is nothing less than passionate. This is the reason that he is willing to use the clean and pure energy that is produced through the interactions and fusion of life, work, and religious practice, in exchange of awaken the quality of the interior space.

1. Moving along with the route to feel the complement of the building and the environment. 2.3. The openness of the layout allows the freedom to feel the pulse of the nature. 4. The water pool calms also enrich the visual sensation. 5. Simple but powerful lines empower the building.

1

2

3

4

5

1. The front water pool reproduces the nature from time to time, and becomes an essential part of life. 2. The wall of waterfall is the symbolic the origins of wind, water, and life. 3.4. Nature is embedded in the materials, and can be seen with one glance. 5. Down to the earth in order to respect the nature environment.

1

2

3

4

1. No unnecessary ornaments. The simple materials and the natural greenery coordinate and support each other. 2. Ground-high furniture in contrast with the continuity of the high-ceiling. 3.4. The turns on the trail allow the view to be appreciated from different angles.

2

4

Philosophy of Nature

What I understood and what I am practicing in architecture is my reactions to the realizations of every existing object and moment. It is like how ancient people behaved when constructing a building and living within. They constantly sought and learned the tangible and intangible power in the man's world that exist in the inner and external cultural system – how to respond to the powerful universe. That is, to reexamine existing rules and orders in the universe from construction, move-in, to settlement, or to re-translate the existing system that is invisible.

2

3

1.3.4.5. The narrow but long layout increases the opportunity to embrace the nature. 2. The spirit of the interior space depends on its relationship with the natural rules.

The process of re-translation keeps the value of spirituality and divinity in architectural construction, and is my way of learning under the traditional religious ceremony. It is also a process of learning, practicing, realizing, and unifying when I feel lost. To practice architecture or to let architecture practice has led me to learn and realize. In the journey of learning and realizing, we accomplished the ceremonial services and guide people to follow.

1. Unpolished log pillars bring the building into the nature. 2. The glass walls assure the view is unblocked. The refractions provide the interests of multi layers. 3. The cozy warm and elegant interior space is surrounded by nature. 4.5. The beauty of the lights and shade are touchable indoors and out during the day.

2

3

1

4

5

1

2

3

4

5

6

1.2.3.4.5.6. Divinity exists wherever light shines. It is eye-catching and moving. 7.8.9. A garden on the first floor and a balcony on the second floor, the natural beauty become visible year round.

An Aesthetician in Lov...

IQ INTERIOR DESIGN CONSTRUCTION CO.,LTD

Living is a simple and life-long practice. Aesthetics are comfortable and eternal. Dong Lin Interior Design Construction has been endeavoring to join these two parallel lines together to sketch elegant lines and three-dimensional background in every building. Then, the designer would like to awake the charm of the space with the decorations of furniture, light, and shadow. Constructing a space for everyday life, it allows users feel unusual happiness in daily life.

Text: Yves

Photo provided: IQ INTERIOR DESIGN CONSTRUCTION CO.,LTD.

JOHN LIAO

1996 Established IQ INTERIOR DESIGN CONSTRUCTION CO.,LTD.

Rewards

2011 DECO Magzine Top Design

2011 TID AWARD for Living Space Design — Single-deck

2012 TID AWARD for Living Space Design — Single-deck

2012 TID AWARD for Living Space Design — Duplex

2012 TID AWARD for Commercial Space

2012 TID AWARD Shortlisted for Office Space

Publications

IQ INTERIOR DESIGN COLLECTION, IQ INTERIORS QUARRY

DESIGNER STUDIO

Top Work Industry Factory & Office Complex A Garden Architecture

An office building gives people the feeling of being reliable and secure. Nature opens up all the sensual experiences. This office building is designed as a supreme size, three-dimensional, and non-column suspension structure. It sculpts a stable structure and visual tension. With the thinking of"interior landscape," Dong Lin Design brings garden landscape around the architecture to build up the intimate relationships between the building and nature.

1. The stone facade of the building is a natural painting of light and shadow. 2. Surrounded by green trees, the architectural structure appears to be light and vivid, not heavy at all. 3. With the help of lightweight materials, the indoor space is given the layering of depth of field.

2

3

1

When people raise their heads, they see the façade made with simple stones with the texture of fair-faced concrete. It offers an ideal backdrop for the green plants with either front-lighting or back-lighting. The beauty of the texture and the projection of the kaleidoscopic shadows of the trees enhance the 3-D feel of the building and the green plants in the garden. Displaying corporate culture, it also communicates the charm of life force.

1.2. Accompanied with the surrounding landscape in the garden, it adds more layering to the space. 3. The indoor space and outdoor space are connected to create a smooth view.

Regarding the interior space, the garden landscape is still an indispensable part of the view. In the informal spatial structure, it is possible to appreciate the spatial sublimity along the routes through the steel's lightness and glass's transparency from different perspective and depth of fields.

1. The use of lightweight materials diminishes the weight of the spatial structure. 2. The high glass shortens the distance between indoor and outdoor. Through the curtains, the light is sieved to create a comfortable atmosphere. 3.The stone's textures communicate diverse emotions. 4. The layout is done with irregular cutting, allowing the routes and views connect with the landscape outdoor. 5.6. Different from common stairs that communicate the feeling of confinement, the iron and glass orchestrate the layering of the beauty of field of depth.

1

2

3

4

5

6

The Twin City House The Longing for an Elegant and Warm Space

Jennifer Tang said, "The point of origin is interpreted with classical charm with white and modest facades installed with sculpted plates, on which it depicts a quality but not noisy spatial background." Matching with fine furniture, decorations, and lighting it makes the views harmonious and comfortable, highlight the spatial values and bring sublimity to everyday life.

1. The facade made with plates with classical simple lines adds to the elegance and tranquility of the garden. 2. The design is exploring the spatial feelings and values. 3.4.5. The space returns to the point of origin. The accompanying objects bring sublimity in a ration milieu.

1. In the white backdrop, the black lines extend the smoothness of the vision. 2. Light reflects the spatial vividness of life.
3.4. The classical filament plates construct elegant visual rhythm along the curves.

The structure is spacious. Captured with lines, the beauty of the contour is accentuated. It further enhances the three-dimensional functions and configurations. Each floor is saturated with the female owner's longing for elegance and charm, without obstructing her independence. It also recalls the male owner's expectation for grandeur.

1. With the details of the lines, it promotes quality and taste. 2. The decorations highlight the space to create grandeur. 3. Like people, the people are quiet and charming.

Supreme The Grandeur of Dimensions

Dimension is always the key to define the frandeur and values of a space. This 5-level house has the predominant advantages of capacity. Dong Lin Design abandoned the traditional design of terrace house. Reorganizing the use of the order of space, it creates not only visual tension with horizontal plane but also the diverse layout of vertical lines.

1. Through the opening of the layout, the greenery outdoor is also the focus indoors. 2. The use of black and white creates a classical tone.

1

2

1

2

3

4

Especially in the process of completing its functions, it connects the landscape outside with viewers inside, decorated with huge pieces of furniture. Of course, space does not necessarily imply magnificence. It is important to attend to the details to make the design concrete. The owner has considerable knowledge in design, with insistence and high demand on beauty. The designer demands highly on the texture, tincture, and the visual harmony of the structure on overall design to provide users with a living space of chic and style, creates a huge entrance hall, huge living room, and huge bedroom in every floor, giving users maximum comfort and magnificence in a huge area.

2.3.4. In the grand space, the details are carefully done to communicate smoothness and warmth.
1.5.6. In the huge space, the façade made with different materials communicate different situations.

5

6

Yogi Daichien American Elegance

In the realm of beauty, Dong Lin / IQ Interior Design Construction Co. Ltd insists on elegance and taste in its design. With ample thinking in lifestyle, it is able to construct unique spatial syntax. This house displays a magnificent style created with a continuous facade to generate a smooth and tranquil atmosphere. According to the owner's expectation, the space can satisfy multifunctional uses; therefore, the designer utilizes the space practically and neatly. Configuring the objects of everyday life tactfully, she delineates the routes and order in the huge area wisely, with no redundancy so that the users enjoy a broader view and more leisure atmosphere.

1. Elegance originates from the familiarization of the classics. 2. In the bright white tone, the dark wooden floor in the bedroom creates a serene feeling. 3. In addition to pure white, the touch of material itself communicates rich visual layering.

2

1

3

Of course, there are diverse emotional perspectives in everyday life; therefore, in the bright atmosphere the furniture, decorations, and lighting are either in black, grey, or silver to avoid visual fatigue. With harmonious and rice color proportion, it injects humanistic profoundness and fashion into the space of everyday life to interpret the mature America style without any cliché.

1. The symmetric design enhances the existing spatial grandeur. 2.3. Simple and harmonious furniture arrangement. 4. The integration of functional demand and aesthetic design creates a grand, smooth, and comfortable atmosphere.

2

3

4

1

2

1. The introduction of grey adds depth to the white. 2. The black backdrop makes white more charming. 3.4.5. The depiction of the curves softens the view. 6. The square with a circle placed inside displays a sense of generosity.

3

4

F. SCOTT FITZ-GERALD
THE GREAT GATSBY
F. SCOTT FITZ-GERALD
THE GREAT GATSBY
GEORGE ELIOT
MIDDLEMARCH
EVERYMAN'S LIBRARY
FORD MADOX FORD
THE GOOD SOLDIER
EVERYMAN'S LIBRARY

5

6

American Style Architecture / Home Design

Dong Lin/IQ Interior Design Construction Co. Ltd is specialized in designing American style architecture. The designer hopes people change their mindscape when they are moving around the landscape. In addition, they can be inspired to have new visions in life.

1

2

3

For example, this business building is design with the feel of magnificence and elegant sensation. Demanding on the quality of the furniture, the designer has the indoor decoration appeal to the harmony between the space and the people. Therefore, there is not neglected or idle corner. When the occupants return home, they love to interact with the space to explore their life force.

1.2.3. Integrating Victorian style and classical American style, it redefines the possibilities of office space. 4.5. With the decoration of home space elements, the commercial space becomes more charming. 6. With the windows opened suitably, it creates an interesting extension of view.

4

5

6

Complying with the idea of "design for nature with no artificial space", the spatial configuration is saturated with meaning and value. Blending Victorian style with American classical style, the designer uses symmetry and depth of field wisely to redefine the commercial space and elegant spatial aesthetics with the changes of light and shadow in the smooth flow of the space.

1.2.3. Sculpted by the light, each line in the space display extraordinary charm. 4. No exaggerating and no fancy design creates agility and elegance. 5. The continuous extension in vertical or horizontal manner further enhances the spaciousness and openness of the layout. 6. Controlled at will, the layers become clearer.

1

2

3

4

5

6

1

1.2.3. In each corner, we discover the beauty of layering in the frame. 4. Walking in between the non-exaggerating design, we feel saturated with energy. 5. Like home, the placement of the displayed items is done upon the harmonious relationships between people and space.

A Creator Well-Versed in Literature

CHING-CHI ARCHITECTURE & INTERIOR DESIGN ASSOCIATE / SU, JING-CI

Narrated from the literary perspectives, the artistic features hidden in modern architecture came into being. Different from the construction with cold lines, the architecture depicted by the architect Su Ching Chi displays diverse meanings and connotations. In the performance of each case, he emphasizes "the existence of architecture," which is an extension of accumulated memories, extension of realms, or a refraction of natural scenery. Transforming architecture into artwork, he gives them a lofty life force. In Taichung, a land rich in creativity lives with the nutrients and cultivation of humanistic architectures.

Text: Yves

Photo provided: CHING-CHI ARCHITECTURE & INTERIOR DESIGN ASSOCIATE

SU, JING-CI

1997 Established Yuan Jing Architecture & Interior Design Studio

2003 Established Su Jingci Architecture & Interior Design Studio

2006 Established CHING-CHI ARCHITECTURE & INTERIOR DESIGN ASSOCIATE

Now the Visiting Lecturer of Feng Chia University

Rewards

2009 TID AWARD for Temporary Building Design

2010 TID AWARD - Gold

2011 TDC Golden Pin Design Award

2011 TID AWARD for Public Space Design

2012 TID AWARD Public Space Design

2012 TID AWARD - Gold

1

Memory Lihe Construction Co. Ltd. and Poetic Public Construction

Ching Chi Design had been experimenting with the experience and display of the design of spatial "existence."Normally, most designers employ frequently used elements such as walls, water, trees, and more. These are the childhood spatial memory wandering in people's minds. In this project, the architect designed the entrance of the community at the end of the base to create the wandering experience. Then, he designed a water pool between two buildings. Behind the camphor tree forest, he used wall frames to expose the water yard in the environment around to entertain the neighborhood. The wall symbol was a dedication of memory to the neighborhood. The floating platforms on the water yard interacted with the users with the vegetation and fish in the pond during the daytime, becoming the center stage the community. With a water domain that could not be trespassed on, it created an ephemeral"air wall" between the community and the neighborhood. At night, the light of the water yard was turned down intentionally to maintain the sound of the water flow to create a black domain of water. The crisscrossing illuminated hallway became the image of dim light that symbolized protection and home. In storms, heavy rain splashed the water yard. The sound of the carp hiding under the lotus leaves recalled a memory from afar.

1.2. With multi-layers and borrowed landscape, it employs the glass-box like hallway in its spatial layout to illuminate the landscapes mutually; employing the difference in light control between day and night, it creates special visual effects for the space. 3.4 The fountain in the yard is the center of the community; with a water domain difficult to cross over, it maintains an "invisible air wall" between the community and its neighborhood.

2

3

4

A transparent glass hallway connected two building, in which six indoor and two outdoor zones were designed in the public space. The high and low, front and rear glass boxes interacted with each other, becoming the backdrop and front view. The different light sources during the day and at night were controlled "in-between" to create contrast luminance, resulting in spatial and visual fun. In addition, all the materials, including the layering rocks on the walls, the stone plates on the ground outdoor, the texture of the stone floor indoor, glass panes, rain shield plates, wood cupboard's ratio, the theme and ratio of photos, the theme of sculptures, and more were all designed with the ratio of longitudinal lines to create a unified texture, allowing the space to offer the experience of concealment. At last, different behavior or experience may come to a different result. And in the process, the emotions that is triggered by different angle, different lighting, different shadow or different weather has formed the decisive variables.

2

3

1.3.The proportions of all the materials, including the laminated stone walls extending from indoor to outdoor, the fine texture of the stone floors, the glass wall frames, rain shades, the canvas offset of the photos are stretched to create a unified texture, allowing the space to conceal the overall experiences. 2. Create the authentic spatial experience, such as the library, in which simple drinks are served in addition to the loaning of books and magazines.

1

Memory The Reception Center of Lihe Original Landscape

The project is located at the southeast of the Metropolitan Opera House in the 7th readjustment area in Taichung. It was a temporary reception center of an architectural project. In years of development of design, the theme of "wall" was used in this project, in which a wall plate design was seen elsewhere, including the structure's walls, the walls off the structure, and more. The prototype of this project was adapted from a part of a design done twenty years ago. At first, that design was the entrance of an art museum. Now, it developed into a building structure. Although they resembled each other in form, they were different spatially. It attempted to design an image that could represent the construction company. Attempting to extend the center's space without display anything, in fact, it was the theme of the exhibition of space.

1. The creative idea of the lobby is inspired from the endless memories that resemble an infinite rail track. 2. Create special effects for the concrete walls, as if growing up from the ground. 3.To develop and demonstrate the theme of "space"; the main structure is constructed with faced concrete; employing special horizontal shading panel, it displays the transparency of space.

In its appearance, the concrete building structure impressed people like sculptures when it was finshed, like ceramic crafts in a building. In the space, an interface design was placed to disrupt the inner and outer space. Extending the pool indoor, glass was used to seal off the water, blurring the lines between indoor and outdoor. Through the windows of the major structure, daylight was let in intentionally. Meanwhile, the metal fences were designed with careful calculation to shield the afternoon sun. Likewise, there was no need to install any shade in front of the building to maintain its transparent atmosphere. In addition, the slanting straight lines were compressed into space by the walls that were pressing down to create a dynamic spatial feel.

1. Employ concrete on the architecture's façade to create the sensations of art works. 2. Transgress the inter- and outer-space with the special interface design, allowing the pool to extend indoor to blur the lines between indoor and outdoor. 3. Allow the architecture's line continue to the indoor space, creating the visual effects of have sculptures elsewhere. 4. The interior space continues the feel of simplicity and conciseness to create tranquil effects.

1

2

3

4

1

2

Broad Treasure the Public Space of the Lotus Garden

With simple horizontal lines, the tones of the color increased as the entrance extended, employing the compression method. It led visitors to enter a bright and transparent space. It borrowed the view from the bright and broad space in the end. Finally, with the thinking of walking into a broad space and the landscape in scholars' paintings in the past, it allowed people to experience the changes of the four seasons in the modern transparent pavilion. This project won the Golden Award for Public Space in the TID Award in 2010 and the TDC Golden Pin Design Award in 2011.

1. The lines of the facades display agility and lightness. 2. The landscape opens in front of our eyes, creating a landscape within landscape. 3. The mirror pool along the trail reflects the changes of the landscape.4.This transparency shines even at night.

3

4

1. The brightness of the glass pavilion constructs the taste and feel of the garden's charm. 2. A patch of green, a corner of stones, and a pond of water communicates humanistic atmosphere in nature. 3.6. The interface between indoor and outdoor, regardless of size, it displays extensiveness and complacence. 4. The indoor space delineates stables and the beauty of the sculpted details. 5. The combination of warm wood and cold glass creates harmony.

1

2

3

4

5

6

Memory the Public Space in Original Landscape

About four years ago, the designer completed a project of a reception center. With the display of "space" as the core issue, the main structure was built with fair faced concrete. With a special sun screening method, it displayed the transparency of the space. The construction of the building underwent the process of planning and building at the moment. It extended the little impression of "existence" from a scene at the backyard of a small church, the chimes of the bells under the roof of an Eastern building, the happiness of seeing smoke coming from the chimney of a small cottage in a foggy mountain in the early winter. Transforming such emotion, it was projected onto the display of space.

1. Construct a sky garden on the building's roof. 2. Display sculptures in the landscape to create magnificence.

1

2

1

2

3

Such space was constituted by the image of an endless rail track (the hallway in the lobby), the reflections of the water(the pool on the first floor), and1feeling of growing from the land (the fair faced concrete walls), the light in a dark night (the dim light in the dark), wandering in a forest, and the sound of the branches strummed by the wind (the camphor trees and the osmanthus flowers woods on the roof).

1. In the pool area, the reflection is used to reflect the pictures on the ceiling to create a special view. 2. Illuminate the black hallway with indirect light to create a crisscrossing picture. 3. Create small forests in the city with green waterscape.

The Association of Black the Public Space of HuiYu's Feng Tower

This project was mainly constituted with five basic elements: 1.The warm and bright color light in dark tones was extended from black stones inmost of the space. Decorated with few materials (stone and wood) of warm tones to match the black diffusion at the bottom and the lighting on the light colored materials, it created a hidden dramatic contrast. 2. Simple elements were used. With board-like elements of various sizes, it displayed the beauty of pureness in the space constituted by pure elements (eaves and pillar walls). 3. A utopian small forest was installed in the city with dim light. With its reflections on the swimming pool, an artificial natural landscape was displayed.

1. The architecture itself is built with black stones, extending to most of the area in the space; matching with diffusion and projection of the black backdrop and the projection of the light color materials in the area to create the planned dramatic contrast.2.Walking into the architecture, it shapes the secret forest created by the forest and its reflection on the pool.

2

4. With the transparent and symmetric space, it fused with the city. With the sculptured corner in the city, it integrated with its neighborhood. At night, the lobby was transparent and illuminating, like a huge lantern. In the midst of the night view surrounded with the urban greens, it added a kind of the beauty of sadness to the city. 5. Existence kept crawling in the dim world. The designer attempted to search for floating but warm pictures to rely upon. It might be the feeling that could be felt but not grasped. It reminded us that in the dim light at dawn, it is possible for us to experience of the beauty of life in that very moment. All the rest were reflections of a dim light in the dark.

1.5. Reflections of the greens on the waterscape create the feel of a small utopia. 2. Discarding the magnificence, it replaces the spatial depiction with pure elements. 3. In addition to the demanding on architectural materials, all the decorations have become the highlights of installation arts. 4. Employing open methods, it creates a magnificent night view with the tension between the architecture and the city.

1

2

3

4

5

The Association of Stone Hedge the Reception Center of the Park Tower by Yuan Lin Construction

The pure white oval space of the reception center was surrounded by primitive stone hedges. This old but new image attempted to inspire the sensation of existence. The base of the reception center (the future construction site of the project) was located by the Taichung Amphitheatre Park. On the white oval architecture, the frame walls resembling the stone hedges were erected. On their surface, faux rusted lead and iron paint was sprayed. The building was situated in the middle of a pool. The gate of this building was opened at a side of its back to offer the front view of the park. In addition, it also exhibited art works in the future square to the pedestrians. At the other side at the back, there was an entrance leading into a white cave. In order to present the future impression of pureness, four electric controlled huge walls are installed at the four entrances as gates to constitute the image of floating space. Between the walls, stairs leading to the second floor were built. In order to construct a better view, the sample house was constructed on the second floor. Likewise, visitors could have a full view of the Amphitheatre and the park, gazing through the stone hedges. When night fell, due to the light and shadow of the double-walls, reflections were formed. It furthered created the flow of wind between the walls. Visitors could immediately feel the existence of time in this temporary building. The designer expected the visitors to feel the complexity of this"virtual time."

1. In the midst of white space, a huge area is preserved as empty space to motivate more thinking. 2. In the bedroom's space, pictures are used as visual emphasis to create an artistic feel . 3. The façade of the architecture employs the stone-hedge design of the ancient times to create layering. Recalling the waterscape, it extends a natural feel. 4.5.6.7. Entering the interior space, the swirling stairs give people a spatial feeling of entering a stone cave.

4

5

6

7

Lotus Garden

Creating contrasting highlights with the water, garden, pond, and landscape, it constitutes a leisure space and brings tranquil environment interesting views.

1. The water flowing image display life force. 2. The decoration of light adds tranquility and stability. 3. The arrangement of each tree aims to enrich the emotional realm. 4. The marble and mirror concentrate into shinning spots with indirect light. 5. The sculpted leaf-shape lighting on the ceiling of the reception area harmonizes with a leisure tone. 6. The deep color accompanies the brightness and extension of the outdoor landscape.

4

5

1

HsinYeh Macro Vision

The main entrance is located on a minor road. On the side of the major road, a tranquil green garden was design in answer to the noise and the bustle of the business district, to become a refuge in the noisy city. There is no wall in front of the building. With the decorations of the sculptures and pond, it allows pedestrians see the community directly. With the layering wall and extending roof and the contrast of vertical and horizontal lines of gradual differences, it creates qausi-Chinese garden spatial layers.

2

1. The semi-open space allows users enjoy the ensemble of the sun, fresh air, and water. 2. With tactful juxtaposition, it creates spatial layering. 3.5.6 In the public area the extension of the wall and the floor surprises people with an amazing vision. 4. Because of the night, it highlights the tranquility and elegance of the space.

1.2. The profound delineation of the lights demonstrates infinite visual charm. 3. The arraignment of the landscape harmonizes with the architecture. 4. The corridors are not only routes but also a buffer between indoor and out. 5. The serenity and tranquility indoor answers the noise and chaos outdoor.

1

2

3

4

5

Initiator of Impressionist GuanZhi Trend, the Pursuit of Ultimate Perfection

CTH DESIGN ASSOCIATE / CHENG, TANG-HUANG

Designer Cheng Tang-huang was born in 1965; he established his architect and interior design studio in 1989 and remains the chief designer today. He has a wealth of experience and specializes in the application of materials and light to express spatial emotion. He is passionate about the dialogue between the gradation of lights and one's spirit. He believes the spatial dialogue is generated from the moment the heart is touched, and that is also the moment his design begins.

Text: Sofia

Photo provided: CTH DESIGN ASSOCIATE

CHENG, TANG-HUANG
1965 Born in Zhanghua, Taiwan
1989 Established Ancient Greece Design Studio
1999 Established CTH DESIGN ASSOCIATE
Graduated from Institute of Architecture, Feng Chia University
Now the Principal of Straight Line & Broken Line Design Studio

Onsen Papawaqa

ONSEN PAPAWAQA has introduced a breath of fresh air to the design of hot spring hotels in Taiwan. Drawing inspiration from the serene landscape of Taian Township, the designer creates the architecture in accordance with the natural environment; many of the minute details originate from the accumulated experiences of aesthetics in life to form the general theme. The ingenious spatial arrangement is also natural, while the changing light is created through our in-depth understanding of the overall physical environment. We have transformed it into a game, where happiness is ubiquitous.

Designer Tang Huang Cheng, ingeniously incorporates the overall design of Onsen Papawaqa into the natural landscape, embedding it amongst the mountains and rivers, unleashing an unprecedented design trend for hot spring hotels in Taiwan.

1

2

The fusion of illusion and reality on the water permeates like music in the atmosphere and fills the pure environment with touching poetry. The meticulous design not only blends the interior with the outdoor sunlight and greenery, the unassuming precast concrete wall surfaces also creates unique light transformations and balanced playfulness of nature. The astonishing lights and colors also become conceptual challenges for other designs of mine.

1.The combination of such materials as concrete slab, timber and stone used in accordance with nature's existing state in the environment is a humble gesture of respect for nature, as well as a creation rooted in the local and the nature. 2. All procession and pauses in the space are clues derived from nature; all alteration of light and shadows are from an understanding of the entire physical environment prior to controlling and maneuvering the design. 3.Onsen Papawaqa, located in Miaoli County's Taian Township, is 500 meters above sea level with beautiful natural surroundings, endowed with a rich ecology, and distinct fourseasons. The hot spring is sodium hydrogen carbonate-based and possesses considerable tourism potential.

1

2

1. Inside the hall, above the mirror-like water, the virtual and realistic reflections and ripples are crisscrossing each other as if to compose a symphony permeating and expanding, along with the evaporation in the air, to the beat of a pure, poetic and lively dance. 2.To still one's mind in the tranquil forest and to allow the water and breeze to caress you, the tired traveler will be soothed and restored to a inner natural state. 3.The term, Guanzhi ('the pursuit of the ultimate perfection'), corresponds to the engrained beauty of the natural landscape in Taian Township and is revered as the guideline to nurture the visitors' body, mind and soul via the pristine warm water from nature. 4 Connecting the interior and exterior space, arranging for the sunshine and green shades to roam freely in the space, man's pace of living will thus be playfully in sync with nature.

3

4

1

Wein Restaurant and Lounge

Tang-Huang Cheng want to create a new dining cultural space through the marriage of a restaurant and lounge in the seventh district in Taichung, which is an important urban hub of the city. From architecture to interior design, this becomes a substantial, black entity, just like a facet of a gemstone after it is cut. Externally, it reflects the expression of the city. Internally the irregular folded plate and elaborate decoration mold the unique sentiments of the inside. Contrasting phenomena such as tangible/intangible, closed/open, conceptual/abstract, venue/behavior is used to mimic nature in the urban environment. The objective is to allow users to enjoy gourmet food in a leisurely, peaceful corner within the bustling city.

1.2.3.4. It's almost impossible to imagine such a surrealistic transcendental place surrounded by bamboo forests in amidst of the 7th urban re-planning zone. The Restaurant and Lounge Bar are combined for a brand-new dining experience that fuses cultural characteristics with modernistic aesthetics for an innovative cuisine culture, thus marking a new chapter for dining space design.

1."Lines" are omnipresent in our everyday life; whether in architectures, furniture, dishes and so on, "lines" construct space, patterns and texture, shaping various forms of space into various arrangements, arriving at multifarious end effects. 2.3. Irregularly-folded boards show a heavy-loaded wrapping that gives the contours to the interior space and makes a bold attempt at imitating nature, to transpose a stolen moment of peace and quiet for the ruffled city dwellers in their dining experience that is made more enjoyable, pure and simple. 4.The concept for "WEIN Restaurant & Lounge" is of line-based dimensions of various diagonal angles folded to form a solid black mass. Much like a diamond, it undergoes several precise cuttings to reflect the expressions and contours of a city, presenting the frames of human dwellings.

1

2

3

4

Karuizawa

Karuisawa is a renowned chain restaurant. I want to complement the food's original flavor and pure delicacy with an understated, minimalistic exterior design and interior space, immersed in warm, brown color. Ingenious application of lighting design reflects the glass-like lucidity of the room. In the dining area, a metal mesh frame is used as the partition, which creates an entirely different environment compared to the transparency of glass. Besides being fashionable and lightweight, the simple lighting illuminated on the mesh creates different playful patterns. Candlelight from the ironware mounted on the wooden frame brings patrons back to a more primordial time and space.

1.2.3. The deep-sweeping simplistic undertone presents a certain human interaction with food akin to the first tug at the heart string in that instant when humans come into contact with space; the simple beautiful taste of food is thus heightened. The cool-colored undertone as if merely for visual illustration brings the folks - that are enveloped in dark brown colors and soaked in an ambience of warm yellow lighting – straight to a state of brisk readiness with no compromises. 4. The dining area is petitioned off with hollowed frames woven with metal wire and netting. The holes in the nets and the transparency of the glass help with space division without physically compartmentalizing the space, with each divided area expressing different states of emotions by different users. Tired of the tiresome grind of city life, more and more people are attracted to a simple, rustic spatial design for their living space as a gesture to express the desire and the trend to return to nature.

1

3

2

4

Reception Center of The Shining Chunqing

I do not regard sales centers of construction firms as a product design; as far as I am concerned, this is a place where people interact and socialize, therefore I try to simplify every element of the design to present a wonderful living space through innovative philosophies. I employ unpretentious, perceptual design to congeal emotions and repeatedly research every detail. Concentrating on the flow of people in the space, I introduce life into the space as the foundation of my design.

More and more people now want nothing but simple and natural when it comes to interior design. When tired of busy city life, people simply want to get away from it and embrace the nature when going back home.

For The Shining Chunqing, white blocks are used as the framework to build the spatial entity, while succinct lines and perfectly proportioned segments are applied to transitional spaces such as public paths and corner landscaping, so as to join and transform various environments. The same concept extends from indoors to the exterior, where natural light and ambient lighting combine to create a comfortable spatial aura and taste. At night, the architecture reflects off the water surface and emanates the charisma, happiness and warmth of home.

1

1.2. Breezes penetrating the building bring pleasant coolness that needs not any excessive equipment but allows you to experience the beauty of easy living and gentle contact with nature. Also the absence of obstruction leads one's eye line towards the tree branch tips and towards the sky, seamlessly altering one's state of mind. In this confined finite space, infinite imagination s sparked and spurred. 3. he white frame of a modernist vocabulary comes to shape the elegant white pagoda that brings image reflections to the ind of the viewers by the pool; the sunshine, green trees, blue sky, and pool water together pushes the enjoyment up a notch.

2

1

Private Residence

I have always avoided designing a space with excessive décor; rather, I hope to present a discreet, calm spatial thinking pattern and showcase the abundance of life in a subtle manner. Moreover, the extension of view permeates in the space to illustrate unparalleled quality.

1. Sedate and settled nature presents the proprietor as professional, the irregular lines and folded boards forming vivacious vertical planes and semi-transparent texture to lighten and stretch the visual effect that volume masses have on people, while shortening the distance between the common folk and the professionals, prompting for the former to reach out to the latter. 2.3.4. In the mellow interior color tones are injected with fitting interweaving of fine lines and lighting to give that extra push, to highlight the beauty of multilayered conciseness. A number of golden velvet lounge chairs are the manifestation of a poised yet quietly inviting majestic presence.

2

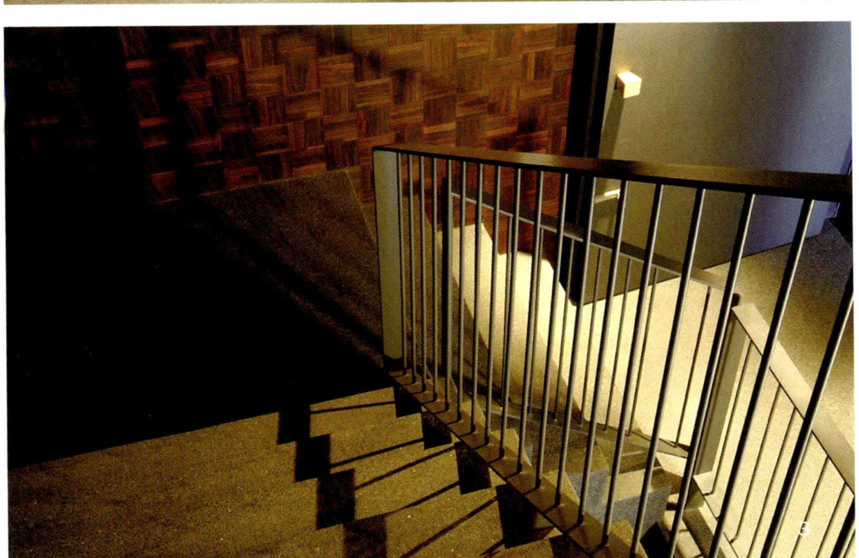

3

4

This is all about returning to the embrace of nature. Consequently, natural aesthetics is applied to form the simple spatial ambiance, and the rich layers of light variation exude an aura of casualness and tranquility. As light reflects off the blank walls, the exquisite, composed aesthetics unintentionally emanate classic cultural and artistic taste. The spatial relationships and layers mingle to create an elegant atmosphere. This is the most touching aspect of life.

1. The urban abodes should also be granted moments of repose, moments of freedom, breaking free from the shackles of concrete jungles, and be granted a seat on a patio to enjoy a pot of tea, to listen to the breeze surfing through the grass. 2. Not to hinge on extensive to patch up an indoor space and not to contritely handle every single need, but to genuinely convert every stimulation into an launching pad for further meditation over what form and what content should the space adopt; only through the construction of time and space can rich layers of nature be hopefully seamlessly presented in life. 3. Fine-tuning human measurements towards nature, towards one streamlined composition, one may be positioned in a man-made living space but will be in one with nature.

1

2

3

1

Form Follows Space

I like simple objects. When giving lectures to fellow architects or students, I generally begin my exploration from the original form of space. Walking out of the space into life is a spiritual journey to nature, where one can appreciate the serene quality of nature and embrace it. When everything magnificent is aroused from within, nature will lead people to treat one another with wonderfulness; such simple quality also inspires my creative thinking. Architectural theory advocates "Form follows function, and function follows form", but Tang-Huang Cheng prefer "Form follows space"

1. The theory of mirror facets reflects a throbbing urban pulse onto a 3-dimensional plane adorned with different expressions at different time of the day. The interface between the measuring body itself and the city is the gentle water and greenness, mutually accommodating as if to embark on a harmonious spas de deux. 2. The introduction of natural light and colors add an extra layer of ease to the interior space; framed from various angles, any given outdoor views can be accessed from indoors.

1

2

3

4

5

1.2. Precise use of lines can be seen in all elements of the design, from standing cabinets, staircases, railings and vertical sections that have manifested the concision and charm of the design. The lines, given all their different thickness and materials, are able to convey, independent of text, to the user the concept of 'shapes derived from space, morphing into utility. 3.4.5. To adroitly use the texture of materials according to the interplay of light and shadow for rich spatial expression is to create a pulsating sensation with layered yet contrasting illumination and darkness. Space begets happiness and initiates exploration in its primal form so that humans can feel for and sniff at the texture of warmth, to retain the true color and original state of nature. One cannot help but reach out and touch space.

1

Papawaqa 22

In order to cultivate a unique sensual experience, the building exterior breaks free from conventional square structure and adopts a folded-board design approach to allow for the volume mass to express a sense of dance-like buoyant playfulness via the outline sculpturing of inward convulsion or outward expansion. In order to convey a low-key yet refined texture, vertical sections of grey-colored stone materials are in place for a rustic charm mixed in with a prevailing feel of elegance, tantalizing and inviting. The interior space is an extension from sleek and concise use of lines, coupled with ingenious link to the structure, to evoke a romantic and chic ambience. Structural designs, such as suspended frames, glass interfaces, extensions of vertical sections, line cuttings, are adopted to create a rich landscape of 'scenes within a scene. In addition, owing to the blurred inner/outer boundary, the line of vision and the line of movement can thus operate on a multi-layered spatial domain freely flowing on either side of privacy and openness.

1. The overall building is surrounded with garden landscaping as if framed by nature to reflect an enchanting yet elusive beauty. 2. The building exterior conveys a sense of playfulness from line sculpturing similar to that of folded boards. 3.4. The vertical section of single-colored, grey stone in vast planes is highlighted with a sense of low-key serene visual effect.

2

3

4

1. J-type guestrooms are like glass boxes where the natural light can stream in through big glass windows. 2. The vertical section of single-colored, grey stone in vast planes is highlighted with a sense of low-key serene visual effect. 3. C-type bathrooms carry a dark color undertone to give a stress-free feel and uses lighting and art pieces for a visual highlight. 4. J-type guestrooms have ceiling-to-floor glass partitions to allow for full visual access to the 'sky well' (refers to an opening in the ceiling to let in sunlight), enticing one to enjoy a light-shadow interplay. 5.D-type guestrooms are internally divided into sleeping, hot spring, rest areas with reclining chairs; such distinct division yet gives off a loose connection within the same room.

1

2

3

4

5

Acknowledgements

We would like to thank all the designers and companies who made significant contributions to the compilation of this book. Without them, this project would not have been possible. We would also like to thank many others whose names did not appear on the credits, but made specific input and support for the project from beginning to end.

Future Editions

If you would like to contribute to the next edition of Artpower, please email us your details to: artpower@artpower.com.cn